Philosophical Observations on the Senses of Vision and Hearing

John Elliott

CAMBRIDGE UNIVERSITY PRESS

Cambridge, New York, Melbourne, Madrid, Cape Town,
Singapore, São Paolo, Delhi, Mexico City

Published in the United States of America by Cambridge University Press, New York

www.cambridge.org
Information on this title: www.cambridge.org/9781108061711

© in this compilation Cambridge University Press 2013

This edition first published 1780
This digitally printed version 2013

ISBN 978-1-108-06171-1 Paperback

CAMBRIDGE LIBRARY COLLECTION

Books of enduring scholarly value

Life Sciences

Until the nineteenth century, the various subjects now known as the life
sciences were regarded either as arcane studies which had little impact
on ordinary daily life, or as a genteel hobby for the leisured classes. The
increasing academic rigour and systematisation brought to the study of
botany, zoology and other disciplines, and their adoption in university
curricula, are reflected in the books reissued in this series.

Philosophical Observations
on the Senses of Vision and Hearing

Although first to suggest the possibility of light frequencies beyond the
visible spectrum, the natural philosopher John Elliott (1747–87) was
better known at his death for his failed suicide in front of the woman
he loved. Tried for attempting to shoot her, he was acquitted but died
in prison awaiting trial on the lesser charge of assault. First published
in 1780, this work was his most important. Contemporary science held
that vibrations of the air were directly communicated to the optic and
auditory nerves and passed on to the sensorium, while Elliot proposed,
through experimentation upon himself, the existence of sensory receptors,
each tuned to only a limited part of the spectrum of physical frequencies.
This insight led him to postulate the existence of what we now know to
be ultraviolet and infrared radiation, thus paving the way for further
discoveries in human sensory perception.

PHILOSOPHICAL
OBSERVATIONS
ON THE
S E N S E S
O F
VISION AND HEARING;

TO WHICH ARE ADDED, A TREATISE

ON HARMONIC SOUNDS,

AND AN ESSAY

ON COMBUSTION

AND

ANIMAL HEAT.

BY J. ELLIOTT, APOTHECARY.

LONDON:

PRINTED FOR J. MURRAY, NO. 32, FLEET-STREET.

M,DCC,LXXX.

T O

SAMUEL FOART SIMMONS, M.D.

F. R. S.

MEMBER OF THE

ROYAL COLLEGE OF PHYSICIANS,

LONDON,

AND ONE OF THE FOREIGN FELLOWS

OF THE ROYAL SOCIETY OF MEDICINE AT

PARIS;

A GENTLEMAN

NOT LESS DISTINGUISHED BY

HIS HUMANITY

AND SUPERIOR SKILL IN HIS PROFESSION,

THAN

BY HIS EXTENSIVE ACQUAINTANCE WITH, AND ZEAL

TO PROMOTE EVERY BRANCH

OF USEFUL KNOWLEDGE.

THE FOLLOWING SHEETS

(AS A SMALL, BUT SINCERE TRIBUTE TO HIS MERIT)

ARE RESPECTFULLY INSCRIBED

BY HIS OBLIGED,

AND DEVOTED HUMBLE SERVANT,

Carnaby Market,
Nov. 4, 1779.

J. ELLIOTT.

a 2

E R R A T A.

Page iv. line 13, for *chin* read *upper lip within the face.*
Page xii. line 4, for *is* read *be.* Page xx. line 7, for *excite*
read *excited.* Page xxiii. line 19, for *at* read *of.* Page lxi.
line 14, for *mean* read *refult.* Page lxxviii. line 1, for *of*
read *for.* Page lxxxviii. line 10, for *inflammable* read *uninflammable.* Page cxli. line 18, for *with the bodies* read *with
bodies.* Page cxix. line 10, after *air,* read *more than by one
another.* Page clxxxv. line 14, omit *more.* Page clxxxvii.
line vi, for *philologifts* read *phyfiologifts.* Page cxci. line 12,
for *walking* read *waking.* Page ccxi. line 2, after *that* read
either atmofpherical or dephlogifticated. See alfo page cxi, &c.

Note. To what is faid in the Appendix relative to the four
principles, may be added, that particles of fire feem to attract
æther more than they do one another, and more than they do
particles of earth. And that particles of phlogifton attract
æther more than they do one another, but lefs than they do
particles of earth. The reafons for this opinion will eafily be
perceived by thofe who read the laft Effay.

PREFACE.

THOSE who have any acquaintance with the Prefs know that the firft fheet of a book, containing the title page, &c. is generally printed laft. This remark may account for fome paffages in the fequel.

THE firft of the following Effays I have had by me many years, and have my reafons for mentioning that it is taken from a folio manufcript containing many other inquiries, which was in the hands of perfons, not my friends, fo long ago as the year 1772; the greateft part of that manufcript was written long before: and the fubftance of the Effay alluded to, has been in the poffeffion of a refpectable philofophical character near three years, though it has not till now been convenient for me to make it public.

IN the fecond Effay, I forgot to mention that the note 1, in the fcale, page 79, for the violin,

may

may be called C; and fo in proportion for the
larger inftruments; or elfe the name may be
varied by tuning higher or lower, as mentioned
in the Effay. The ftrings fitted to the violin
may be very fmall fourths, or good thirds of
the leffer fize. I once fitted a guittar with fix,
and afterwards with eight ftrings, regularly
tuned on the idea of that fcale, and convinced
myfelf that the hint might be profecuted to
advantage on viols. By tuning eight ftrings
according to the mufical intervals of an octave,
the compafs of a fifteenth may be obtained,
without having recourfe to any of the difficult
frets, or lefs melodious notes. And the fame
compafs (which is fufficient for moft performances) may be commanded with only fix ftrings,
by taking in notes above the one fourth fret,
which notes will ftill be good, and not difficult
to hit. Six ftrings can eafily be managed with
the bow : for greater nicety in the execution,
the frets may be marked; and perhaps it will
be found more convenient to ufe the notes on
the two-fifth divifion, inftead of thofe on the
one-fifth, as there will then be no occafion to
fhift the hand. An inftrument of this kind might
either be played by itfelf, or ufed to ftrike in
                                            the

the harmonic notes occafionally in a concert, &c. where their effect would be much finer, than when only drawn from a common inftrument tuned fifths, as would be found on trial. I did not think to infert a fcale for the viol di gamba ; but it may eafily be conftructed by thofe who underftand the theory. It may be remarked, that octave notes *on the fame ftring*, are rather *bearing;* and the like may be obferved of the other intervals : the reafon will be obvious to thofe who read that Effay.

In regard to the laft Effay, I have been a little unfortunate, as a moft excellent treatife has been publifhed on the fubject by Mr. Crawford fince mine was in the prefs *: thofe who have read that truly admirable work, will therefore find fome errors, which otherwife perhaps they would not have known to have been fuch, at leaft for the prefent. My performance may be compared to the firft dawn of twilight ; Mr. Crawford's to meridian funfhine : and it is rather an unlucky circumftance that the *latter*

---

* It was no fault of the Author (as could be made appear were it neceffary) that this work was not publifhed in June laft.

fhould

fhould have appeared before the *former*. Yet it ought to be remembered how highly Des Cartes' philofophy was admired, even by the moft learned, before Newton's Principia appeared. It may be added, that Mr. Crawford had the advantage of being able to profecute the inquiry by experiments; whereas I, having been lefs eligibly fituated in life, could only proceed by mere fpeculation, or guefs-work.

For the fame reafon, I have not been able to pay that attention to the ftyle which is neceffary to a work intended for the Public. The ill-natured critic will afk, why then did I publifh at all. I anfwer, that I fhould not have troubled the world with thefe Effays, if better judges than me had not thought that they contained hints which thofe who have leifure and proper conveniencies might improve into real difcoveries: a fection * in Dr. Prieftley's celebrated Hiftory of Electricity will fufficiently juftify my conduct in this refpect.

* Part IV. fection I.

OBSERVATIONS

ON THE

SENSES.

SECTION I.

OF VISION.

THE ufes of the feveral humours and coats of the eye, and the manner in which light is refracted by them for forming the images of objects on the retina, have already been explained by philofophers. My defign is, to endeavour to fhow in what manner light acts after it arrives at the retina, for caufing the appearance of lumination or colour. This branch of optics has alfo been cultivated; but as a confideration of what happens in this fenfe, will ferve to illuftrate what I have to advance concerning the ear, I fhall enlarge on it, regardlefs of what may already have been written on the fubject, efpecially as my obfervations are of a new kind; or at leaft have not hitherto been fo fully attended to.

It

It has been frequently obferved, that if the corner of the eye be preffed with the finger, a luminous fpot or ring will appear, fomewhat differently coloured; and that from a ftroke on the eye, an appearance like a flafh of fire has fometimes been obferved.

After repeated trials, and putting myfelf to fome pain, I learnt that by preffing the balls of my eyes with my hands, in the direction of their axes, with as much force as I could bear, keeping them fteady, and affifting the preffure with the ftrong compreffion of the lids, and contraction of the neighbouring mufcles, there would, after fome time, appear a large luminous fenfation like a concave hemifphere of light, but not very lucid, and chequered (often in a very regular manner) with dark and lefs lucid intervals. If the preffure be continued, and the eyes winked very ftrongly, the appearance will be much brighter, and will feem to tremble. There will fometimes alfo appear large crooked ftreaks of light, much brighter than the other parts, and with certain vermicular, or eel-like motions. By increafing the preffure till the eyes become quite hot and red, the light will

be

be at the brighteſt, and almoſt as lucid as at
noon day : till this time the appearance is generally of a whitiſh colour, tinctured with
yellow, or orange, like the ſun ; or rather like
the light of the moon, or a candle. By conti-
nuing the extreme preſſure, the brightneſs of
the appearance begins to decay, and the colour
gradually changes from a reddiſh and yellowiſh
white, to a bluiſh one ; and ſometimes ſeveral
kinds of coloured ſpots will appear, as red,
green, blue, and a fine violet which generally
diſappears the laſt, and thoſe more verging to
red ſooneſt ; for now the light totally vaniſhes,
nor can it be recalled by a continuation or in-
creaſe of the preſſure. If now the hands be
removed, and the eyes opened, they will be
quite blind even to the direct light of the ſun ;
and it will be ſome time before they recover
their ſight, and then but by degrees.

THIS experiment is very painful, and it is
not every one that would chooſe to repeat it
after me with the requiſite care. Before I proceed farther, it may be neceſſary to clear up
ſome particulars which are to be met with in
the above account.

A 2   THE

THE luminous appearance feems like a concave hemifphere, or parabolic conoid ; it feems
to furround the face only, and not the whole
body. For example, if I prefs the corner of
either eye, a ring of light will appear on the
fide of the face oppofite thereto ; and by re-
moving the finger at intervals all round the
margin of the eye, the ring will appear to move
all round the face : that is, when the finger is on
the right fide of the eye, the fpot or ring will
appear on the left fide of the face ; and contrariwife. Alfo, when the finger is above the eye,
the fpot appears as if it was about the chin; and
when the finger is under the eye, the fpot appears as if upon the forehead ; but in all thefe
cafes, provided the edge or margin of the retina be preffed, it feems as if it was clofe to the
face. So when the lumination is excited in
the manner above defcribed, the margin of it
feems to touch the face, but the centre of it
appears as at fome diftance from the face, and
the diftance is lefs as you approach towards the
margin; which gives the whole the refemblance
of a concave fegment of a fphere, or parabolic
conoid, as mentioned above. Alfo the two eyes
form but one hemifphere, as I know by caufing
this

this lumination by prefling one eye only. Now
as thefe are affairs (I think) merely mechani-
cal, or refulting from organization, and in
which I could not have been mifled by cuftom
or habit, as is the cafe in many inftances of
common vifion, thefe appearances having occurred at the very firft time of exciting the
above lumination, may they not be made ufe
of to fettle the famous difputes concerning the
inverfion of images in the eye, and feeing fingly
with both eyes? Has not every part of the
retina of one eye an anfwerable part in the
other? Do not the correfponding fibres of the
right fides of both eyes meet in the brain, and
terminate in the left fide of the fenfory? Thofe
in the left fides of both eyes, in the right fide
of the fenfory? and thofe in the upper and
lower parts of the retinæ of both eyes in the
contrary parts of the fenfory, fo as to be in an
inverted fituation in the latter, to what they
are in the former * ?

BUT

* The moft decifive experiment that I have met with againft
the junction of the refpective fibres of both eyes is that blue
and yellow thrown feparately on the anfwerable parts of both

retinæ

BUT it is moft worthy of remark, that by preffing the margin of either eye, the ring appears not as on the oppofite fide of that *eye*, but of the *face*. Alfo the concavity of the luminous appearance which arifeth on preffing the centre of the eye, and its furrounding not the *whole body*, nor merely the *eye*, but the *face*, fhew that the retina in the brain encompaffes the whole of that portion of the fenfory which anfwers to the *face*.

THE luminous appearance which arifeth on preffing the centres of the eyes, as defcribed

retinæ do not caufe a green. The correfponding fibres are not perhaps, united, becaufe then if one nerve was difordered or deftroyed, the other would alfo. They may only run by the fides of each other, or even in contrary directions, and terminate in the brain fo as to form two different furfaces, concentric, and at a fmall diftance from e ch. And this feems to appear by certain circumftances of the chequers in preffing both eyes; in other refpects the opinion feems to be well founded. Thus any two anfwerable parts of the eyes, whether excited by the rays of light, or by preffing the corners or centres, yield but one fenfation, as if but one eye had been affected; excepting that the fenfation is ftronger. In fquinting eyes the ball is perhaps diftorted with refpect to the retina, which may account for any feeming deviation, in thefe inftances, from the above rule.

above,

above, is chequered; that is, fome parts of it are darker than others, and fometimes there appear fpots, and ftreaks which are much brighter than the other parts of the lumination. The caufes of which I take to be that the furface of the retina is not even or fmooth, but has prominencies or ridges anfwerable to the regular form .of the chequers, and which may refult from its ftructure *; fo that the vitreous humour muft prefs on it unequally, and by that means caufe fome parts of the appearance to look brighter, and others darker: for the caufe of the lumination is in the retina, as will prefently be fhewn. The apparent trembling I take to proceed from the trembling of the eye and retina, on account of the violent preffure; the moving vermicular ftreaks may arife from the convulfion of the membranes, or coats of the eye impreffed on the retina, and perhaps alfo from the convulfions and tremblings of thefe parts themfelves; the blindnefs arifeth, I fuppofe, from the univerfal oppreffion of the fibres of the optic nerve expanded in the retina, and this paralyfis of them I take to be the reafon

---

* From its net-like ftructure it derives its name (*Retina*).

why

why the eyes do not recover their fight immediately, but by degrees ; becaufe it muft be
fome time before the nerves can be relieved
from their oppreffion, fo as again to communicate the action of the rays of light on them
freely to the fenfory.

WHEN the corner of the eye is preffed with
the finger in the ufual way, there does not appear an univerfal lumination, as on preffing in
the direction of the optic axis, but only a fmall
ring or fpot of light about half an inch in diametre. There is alfo this difference between
them, that the fpot caufed by the lateral preffure prefently difappears, unlefs the eye be
ftruck repeatedly, or with a quavering motion ;
whereas the lumination excited by preffing the
centres of the eyes, continues without requiring fuch quavering motion. The reafon of
this difference I take to be, that in the firft
cafe only a fmall part of the globe of the eye
is protruded on the retina by the oblique preffure, and of courfe only a fmall fpot, proportional thereto, is excited in the fenfory. The
reafon why it fo foon difappears is, I take it,
becaufe the globe being foft, and preffed but ob-
liquely,

liquely, gives way, and changeth its figure fo as to make the preffure on the retina equal ; whence the partial preffure made by the protrufion on the retina there being removed, the luminous fpot difappears ; but if the finger be lifted up, or the preffure leffened, the globe prefently recovers its figure ; and if ftruck by the finger again, the fame partial protrufion, and luminous fpot arifing therefrom will be caufed, but which will again prefently difappear, by reafon of the globe changing its figure as above ; and hence the neceffity of the quavering motion of the finger to preferve the ring *. But thefe things do not happen on preffing the centre of the eye, becaufe there is then no room for the globe to elude the preffure, which in this cafe is therefore general ; fo that the lumination once excited, continues as long as the preffure re-

---

* By preffing the corner of the eye harder than ufual, I have fometimes excited two rings, one on each fide of the face; that at the fide oppofite to the corner preffed arofe in the ufual way ; the other, I imagine, from the effect of the pref- fure paffing diametrically through the globe, and caufing a protrufion on the oppofite part of the retina.

mains,

mains, or till the nerves become paralytic. And even if the lateral preffure be continued and increafed for fome time after the luminous ring difappears, there will arife an univerfal lumination, as on preffing the centre of the eye with the ball of the hand, and for the fame reafon. But the experiment is fo painful that it cannot be made to advantage.

THESE experiments are beft made in the dark and in bed. Before and after fleep, or fainting, the lumination defcribed above appears in fome meafure even without preffing the eye. Luminous fparkles are alfo perceived by a perfon in a dark room, caufed merely, I fuppofe, by the action of the circulating fluids, &c. on the retina.

Now the general conclufion which I would draw from what has been faid is, " That colours are liable to be excited in the eye which do not at all depend on the rays of light." A conclufion which I imagine no perfon, who properly confiders the experiments, will refufe me.

FROM

FROM the analogy difcovered by Sir Ifaac Newton between the vibrations of the rays of light, and thofe of mufical ftrings, or of the pulfes of air for caufing mufical founds, it is concluded that thefe rays caufe vifion by means of their vibrations; and that the different colours, like notes of mufic, depend on the different times of the vibration: may we not therefore infer, " that fince colours may be excited in the eye, independent of the pulfes of the rays of light, they are caufed by vibrations liable to be excited in the eye, of the fame times as thofe of the rays of light? And that as there are different kinds or tones of colour, there are alfo as many different times of vibration for caufing them?" For the luminous appearance which arifeth on preffing the centre of the eye, is in general white like the fun's light, or rather like that of the moon, or a candle. But fuch a colour as this is found by refracting light to be, not an original or fimple colour, but compounded of others; as red, yellow, green, blue, and violet, with all their intermediate fhades anfwerable to the degrees of refrangibility. And by parity of reafon, fince the

like

like white colour arifeth on preffing the eye,
this alfo cannot be a fimple colour, but one
compofed of the fame ingredients; that it is fo
appears by the experiment above defcribed;
for there fome of the ingredients appeared either feparate, or much lefs compounded. And
even the white varies its colour, being fometimes a reddifh or yellowifh white, and at other
times verging to a blue. The ring or fpot
which appears on preffing the corner of the
eye, is likewife fometimes varioufly coloured,
as others have obferved. Alfo if this ring be
made very luminous, it is tinctured with yellow or red; but if it is faintly excited, it is
rather inclined to a greenifh blue, as is the cafe
in the central preffure. Now, by applying the
old maxim of philofophers, that " Nature does
" nothing in vain," may we not be allowed to
draw the following conclufions? viz. " That
the rays of light could not conveniently be made
to communicate their vibrations immediately
to the nerves, but that the interpofition of
thofe fhewn to exift in the retina was neceffary
to that end. That therefore there are in the
retina different times of vibration liable to be
excited,

excited, anfwerable to the times of vibration of the feveral forts of rays. That any one fort of rays, falling on the eye, excite thofe vibrations, and thofe only which are in unifon with them, not at all affecting the others, and therefore caufe only their proper colour. And that in a mixture of feveral forts of rays, falling on the eye, each fort excites only its unifon vibrations, whence the proper compound colour refults from a mixture of the whole.

N. B. As the following remark relates to vifion, it may be here fubjoined.

MARRIOTTE has publifhed a curious experiment which fhews that there is an infenfible fpot in the retina, at the entrance of the optic nerve. If an object be looked at whofe image occupies the whole furface of the retina, one would imagine from thence that an hole or dark fpot fhould be perceived in the part of the object anfwerable to the infenfible fpot in the retina; but no fuch fpot or hole is feen. I have obferved that images which fall near
that

that ſpot are not perceived as properly defined. In the concave lumination excited by preſſing the centre of the eye alſo, no ſuch ſpot is diſcernible. There is no ſuch ſpot or vacuity therefore in the retina of the ſenſory (if I may be allowed the expreſſion); it ſeems to be filled up by the fibres of the optic nerve diſperſed around the ſpot in the eye : hence the ill defined images there.

SECT.

SECTION II.

OF TASTE, SMELL, AND FEELING.

SINCE the diſcovery of the analogy between colours and ſounds, the various kinds of taſtes and ſmells have been conſidered as ſo many different tones or notes of theſe ſenſations. In what manner liquids and odours act on the organs for the purpoſe of cauſing the reſpective ſenſations, whether by vibration or otherwiſe I pretend not to determine. All that I have to obſerve on the ſubject is, that with regard to taſte and ſmell, the caſe is not the ſame as hath been ſhewn with reſpect to the eye: that is, " There are no innate taſtes or ſmells liable to be excited by preſſing or irritating theſe organs, as is the caſe with the colours in the eye ;" at leaſt I have not been able to diſcover any ſuch by the experiments that I have made : and therefore, if odours and liquids cauſe taſtes and ſmells by means of vibrations, they muſt act immediately on the nerves.

WHETHER

WHETHER feeling, or pain, be caufed by vibration, or in what other manner, I alfo cannot determine, though I fufpect the former. I have only to obferve, that the manner in which this fenfation is ordinarily excited is different from what happens in the eye. Vifion is ufually excited by unifon vibrations, though it may be caufed by preffure or irritation, as hath been fhewn: " but feeling or pain is ordinarily excited by mere irritation or preffure." In which refpect it alfo differs from tafting and fmelling.

N. B. THE fenfes are ufually reckoned but five; I would add thirft, hunger, heat, cold, titillation, &c. But of thefe I have nothing to fay but what may be collected from the obfervations on taftes, fmells, and pain.

SECT.

SECTION III.

OF HEARING.

I HAVE purpofely deferred the confideration of this fenfe till laft, becaufe there are fome particulars of the others which will be proper for illuftrating what I have to advance concerning this. It appears that, in the different fenfes, different modes of operation of the refpective agents take place. In the organs of tafte and fmell, liquids and odours communicate their action immediately to the nerves. In feeling, or pain, nothing but irritation is employed. But vifion, though it may be excited by mere irritation, as hath been been fhewn, and as is the cafe with the fenfe laft mentioned; yet for obtaining the fhapes of objects, and for other purpofes of feeing, it was neceffary to employ the rays of light; which yet do not act, like liquids and odours, immediately on the nerves, but by the mediation of unifon vibrations. It has, I think, been generally fuppofed, that the cafe with the fenfe of

B   hearing

hearing is fimilar to what has been faid concerning tafte and fmell, viz. that the vibrations
of the air are communicated immediately to the
auditory nerve. But that this is not true, may
perhaps be concluded from the following obfervations.

It has been remarked by others, that when a
perfon is fleepy, or tired, when the ears are
fwelled by means of a cold, &c. before and after fleeping, or fainting, in gaping wide, and
on other occafions ; but particularly when the
ears are violently ftruck, a noife is fometimes
heard, called vulgarly " a finging or ringing in
" the ears ;" having obferved it in myfelf, and
particularly twice, when I heard feveral diftinct
mufical founds, I conceived the phenomenon to
be worthy of notice, and proceeded to examine
it. It would be tedious to recount the methods
which I practifed, and the pain I fometimes
put myfelf to in order to find out a method of
exciting thefe founds. Let it fuffice, that by
ftrongly contracting the mufcles on the fides of
my head, and thrufting my fingers into my ears,
preffing different parts of them, more eafy to be
learnt by experiment than defcription, I could

at

at any time excite them, in a confuſed medley, to a great number. And if I made the experiment when warm in bed, and inclined to ſleep, I could at length, merely by preſſing my finger on different parts of my ear, excite ſome of them in a manner ſufficiently diſtinƈt to be conſidered as if ſeparate or alone, and by that means make out a kind of plain tune. And even when they are excited in the moſt confuſed manner, ſome may be attended to and conſidered independent of, or diſtinƈt from, the reſt; as is the caſe when many ſounds are heard together in a concert. By purſuing my inquiry, I found that they had the following properties.

I. THAT theſe ſounds are not audible in the natural ſtate of the ear, but require diſtenſion or preſſure, or, more properly ſpeaking, *irritation* to excite them.

II. THAT the loudneſs or ſtrength of the ſound might be increaſed by increaſing that irritation.

III. THAT the ſame ſound never varies in tone or note;—as I knew by comparing ſeveral

ral of them which I could excite with certainty
at pleafure, with the notes of a fixed mufical
inftrument, with which I at firft found them
in unifon; for I could never afterwards find
any fenfible difference between them.

IV. THAT they do not at all depend on the
pulfes of the air;—for if I excite them ever
fo ftrongly they could not be heard by another
perfon; which, from their loudnefs, muft have
happened had they been the effects of aërial
pulfes.

V. THAT there are a great number of
them.—The exact number, or latitude of their
fcale, I have not yet been able to determine,
becaufe of the difficulty attending the fubject.
It requires practice and patience to excite them
at all to advantage: the ears muft be compreffed by the mufcles around; the mouth occafionally opened or fhut; a gaping raifed; the
fingers preffed not only at different parts of the
hole, but alfo around the external ear * ; the

* By thrufting the finger into the ear, and then withdraw-
ing it, fo as to caufe a vacuum in the manner of a fyphon, the
founds are excited to great advantage, and in this way it was
that I difcovered the low ones.

meatus

meatus auditorius may be filled with water: all thefe circumftances fhould be occafionally varied; but above all, the times moft favourable for exciting them, and a place free from noife fhould be chofen. In fhort, it will be much better learnt by practice than defcription; for I imagine that different means are required in different people, becaufe I find a very great diffimilarity in this refpect between my right and left ear: in the latter I can much more eafily excite them than in the former, neither can I always excite unifon founds in both ears by the fame means. And here it may be proper to remark, that perfons of a light turn of mind, and but fuperficially acquainted with philofophical matters, may ridicule thefe kinds of experiments, and laugh at an attempt to deduce a theory of hearing from " a ringing in the ears." But the cafe, I prefume, will be different with thofe of another caft, who examine the matter with due attention, and reflect, that the moft important difcoveries in philofophy have been fuggefted by the moft trifling and even childifh phenomena. For a long time I could excite no found in my left ear deeper than what anfwered to the middle D of a German flute. I

have

have fince gone as low as B; but in my right ear I can now go full two notes lower, viz. to G. As it is with the utmoft difficulty that thefe low notes can be raifed, it fhould feem that there are others ftill deeper, but which are not excitable by the means above defcribed. Alfo in my left ear, I can raife notes from B to about an octave above, in all the intermediate gradations, or fenfible differences; but from thence, to a great part of another octave, I cannot yet excite them, though ftill higher they may be raifed in great plenty, but in a more confufed manner; and they feem alfo as if they were in a different part of the ear, or more inwardly than the lower ones. From thefe confiderations, it feems to appear that there is a regular gradation of them from the loweft to the higheft; though, on account (I fuppofe) of the particular ftructure of the ear, and my being able to prefs it only externally, I cannot yet excite them. In the right ear, feveral founds, intermediate to thofe juft mentioned, are diftinguifhable; and in both ears I can excite many founds which are evidently unifons to each other. A rumbling noife generally heard in making thefe experiments is not to

be

be confounded with the founds above defcribed, but to be confidered as arifing from a vibration of the internal air, communicated to it by the motion of the tympanum, as will hereafter be defcribed.

Now, from the above experiments does it not appear (analogous to what has been fhewn with regard to the eye) that " there are founds liable to be excited in the ear, which do not at all depend on the pulfes of the air?"

I F this be granted, may we not extend the analogy farther, and reafon in the manner following?

I T is demonftrated by philofophers, that founds are caufed by tremors or vibrations in the air. And therefore, " fince founds may be excited in the ear, which do not at all depend on the pulfes of air, they are caufed by vibrations liable to be excited in the ear, at the fame times as thofe aërial vibrations which caufe the fame founds. Alfo, as there are many different notes or gradations of thefe internal founds, there are as many different vibrations liable to be

excited

excited in the ear for caufing them." The ufes of thefe founds may likewife be prefumed to be analogous to what was fhewn of the innate colours; viz. "That the air, external or internal, could not conveniently be made to communicate its vibrations immediately to the auditory nerve, but that the interpofition of thofe fhewn to exift in the ear, was neceffary to that end. That, therefore, there are in the ear different times of vibration liable to be excited, anfwerable to thofe of the air, for caufing the feveral gradations of found *. That any one time of aërial vibration, acting on the ear, excites only that internal one with which it is in unifon, not at all affecting the others, and therefore caufes only the anfwerable found. And that in a mixture of feveral forts or times of aërial vibration beating on the ear, each fort excites only its unifon vibration, as was fhewn to be the cafe with regard to colours."

---

* The fcale of audible founds is faid to be about eight octaves; and the loweft found is caufed by a vibration at the rate of thirty in a fecond.

SECT.

SECTION IV.

*Concerning the Manner in which we obtain an
Idea of the situation of sounds, and other
Phenomena of*

HEARING.

MY design in this Essay was to establish the doctrine of Internal Sounds, as delivered in the last section. With regard to the subject of this, I confess I have nothing to offer which I can satisfy myself of the truth of by experiment, and would only wish to excite the attention of philosophers to so curious an inquiry.

OBSERVATION I. The sounds which are excited by pressing the ear, are weak in comparison to the loudness with which they may be excited by the vibrations of the air; and yet it is well known, that the sound of a musical string is much more easily excited by striking it, than by an unison vibration. Hence we
find,

find, that, in the ear, proper methods are employed for increasing the force of the pulses
of these vibrations. I pass over what share
the tympanum, the bones, &c. may have in
this intention, and shall only obferve, as others
have before me, that the labyrinth and cochlea
are contrivances fimilar in principle to the very
means which we ufually employ for the purpofe of increafing the loudnefs of founds. One
defign of the ftructure which we obferve in the
ear, therefore, feems to be to increafe the force
of the pulfes of aërial vibrations, the better to
enable them to excite the internal founds.

Obs. II. If the centres of both eyes are preffed, only one concave appearance is formed;
which arifeth from hence, that every part of
the retina of one eye has a correfponding part
in the other, as was fhewn in the firft fection.
but if the ears be preffed, the founds do not
appear to be thus united, but thofe of each ear
are heard as on the refpective fide of the head.
This difference is the more remarkable, becaufe
experience fhews that we hear fingly with both
ears, even as we fee fingly with both eyes.

Obs.

OBs. III. In exciting the innate colours, a wide circular fcene appears, in which, as in the retina, objects may be placed in different fituations with refpect to each other: but nothing anfwerable to this can be obferved on exciting the innate founds; neither does the cochlea or labyrinth appear to be at all adapted to fuch a purpofe,

OBs. IV. Sounds may be excited in the ear by the vibrations of the air though the tympanum and little bones be deftroyed, as hath been obferved by others. The cochlea and labyrinth, therefore, are properly the ear, as the retina is properly the eye; and the tympanum, bones, &c. are only appendages fubfervient to certain conditions of hearing, as the humours of the eye are for feeing,

OBs. V. It has been thought by fome, that aërial founds, from whatever quarter they come, affect the fame parts of the ear, becaufe vibrations in the air are fuppofed to be propagated alike on all fides. We readily judge, however, at firft hearing, from what quarter without us a found comes; and this is fo true, that

(as

(as in optics) a found is heard as in that place or from that quarter from whence it was laft reflected: and when walls, or hills, are duly pofited, the original and the reflected founds are both heard as in their refpective places. So in a concert, we hear many founds diftinctly from each; and not that only, but we judge immediately from what quarters around us the different founds heard at the fame time come. Yet, whoever confiders the ftructure of the cochlea and labyrinth, and the general, confufed, and fimilar manner in which founds muft enter them, from whatever part they arrive, muft own, that they do not feem at all calculated to anfwer to thefe phenomena. After the moft attentive confideration, it feems to me that the tone or note, and the ftrength or loudnefs, are the only parts of hearing with which found is concerned.

OBS. VI. I had often wondered why the malleus fhould be fixed to the tympanum, and that a cord, or nerve fhould run acrofs that membrane behind, becaufe they feemed to me to hinder its vibration. Nature, however, has a reafon for every thing fhe does, and therefore

fore fome purpofe is undoubtedly anfwered by them. It has fince occurred to me, that if the tympanum had been free, like the head of a drum, it would, like that, have been capable of yielding only one found or note at a time *; and therefore for every different note it muft have been proportionally tightened or relaxed. But on the contrary, we can hear feveral, or many different founds at the fame time, and even judge of their fituations, as obferved above ; which perhaps could alfo not have obtained if the tympanum had been free.——Are not aërial pulfes, according as they come from a different quarter, made, by means of the external ear and meatus auditorius, to beat upon a different part or fide of the tympanum ?——Do not the contrivances abovementioned prevent the vibration being uniformly communicated to the whole membrane, and confine it to the part immediately affected ?——Any part being paffive to, or liable to receive, any time of vibration that the air may imprefs on it, and dif-

* Whether this holds with regard to receiving and tranf-mitting of founds from the air, as well as by beating it with a ftick may be doubted. It would only perhaps be done lefs clearly or diftinctly when free, than otherwife.

ferent

ferent vibrations, in different parts, at the fame
time?

Obs. VII. By repeated experiments, I find
that the internal founds are not in the tympanum. When I touch that membrane with a
probe, I find indeed that it has a motion, and
that motion occafions the rumbling noife which
arifeth on exciting the internal founds, as before obferved. But I take it that it occafions
that noife only by acting on the internal air,
and thereby exciting the anfwerable innate
found. This motion I imagine is caufed by an
alternate action of the mufcles of the malleus,
which pull it to and fro, for it is a motion of
the whole membrane: it feems to be performed
but very few times in the courfe of a fecond,
and is always, as far as I can find, about the
fame fwiftnefs, though the ftrength or latitude
of the motion is very variable. I fufpect that
this motion of it is continual in fome degree;
for if a wilk-fhell, or other hollow body be
applied to the ear, a noife is heard like the
waves of the fea; which is a very common
experiment, and feems to proceed from the motion of the air (in confequence of fuch action
of

of the tympanum) reflected and made fenfible by the hollow body fo applied. I repeat, however, that the internal founds are not in the tympanum: neither does it appear to me that the whole tympanum, or the mufcles of the malleus, are by any means capable of alternate action fo fwift as the vibration of aërial founds, fome of which are at the rate of feveral thoufands in a fecond. Confiderations like thefe have led me to fufpect that the motion of the whole tympanum, above defcribed, and of the little bones, have lefs fhare in the bufinefs of found or actual hearing than has been commonly imagined *. Are they not either wholly or chiefly fubfervient to the purpofe of exciting the paffions, and other affections of the mind and body, which we find, by experience, to be the confequences of certain conditions of aërial founds ?

Obs. VIII. The tympanum is fo very fenfible that we cannot well make experiments on

---

* It has been thought that the whole tympanum is put into vibration by every different found; but this will appear abfurd, when two or more founds are acting on it at the fame time.

it. Yet if any one choofes to try, he will find,
by touching it with a probe, that the fenfation
does not feem to be confined to that point, but
to affect the fyftem in a more general manner.
If the motion of the whole tympani of both
ears, defcribed above, be fenfibly excited, efpecially if the ears be clofed, the fenfation feems
to fill or furround the whole head. I had a
patient (to whom I can refer the doubtful)
who appeared, by the fymptoms, to have had a
fuppuration in the barrel of the right ear; for
putrid matter was afterwards difcharged from
thence into the mouth. If this patient leaned
her head forward, fhe felt, as fhe expreffed
herfelf, *her brains fall forward,* and if afterwards fhe held it backwards, fhe thought fhe
felt *her brains fall backwards again;* which
made her fancy that her brain was loofe in her
fkull; but if fhe lifted her head up, by placing
her fingers about that ear in a particular manner, thofe things did not happen; and thence
fhe thought her brain, during that time, was
reftored to its right place. From the laft-men-
tioned circumftance, the caufe of this feeming
affection of the brain was undoubtedly in the
ear; perhaps one of the little bones had been
loofened

loofened by the fuppuration, or matter may
have floated in the cavity : hence, according
as her head was moved forward or backward,
it fell againſt the fore or back part of the tympanum, or elſe of the barrel itſelf, but which
motion was prevented by placing her fingers
and lifting her head in the manner defcribed.
Theſe phenomena ſeem to indicate that the
nerves which ſerve either the tympanum or barrel for the ſenſe of feeling, are ſo difpoſed in the
fenſory or brain, that if the organ be affected
in one point, the ſenſation ſhall be felt, not as
in the part affected, but as in the fore part of
the head. If in another part, it ſhall be felt as
in the back part of the head : and perhaps there
are other points of that organ which corre-
ſpond with the whole ſurface of the head re-
ſpectively. If this be the caſe, then, if the air
beats upon a certain point of the tympanum, a
found ſhall be excited in the labyrinth, and at
the ſame time a tremulous ſenſation of feeling,
of the ſame degree of ſwiftneſs, ſhall be excited in the fore part of the fenſory or head, and
thereby give us the idea of the found coming
from the front; for the ſenſe of feeling affects
us more powerfully, or is more intimate to us

C   than

than that of found; and therefore the attention of the mind is chiefly engaged by the former, and thence tranflates, as it were, the
found itfelf thither ; or rather, the ideas are
affociated by cuftom or habit. To give an
inftance of this fuperior power of feeling :
though the fhapes of objects are painted in the
retina, yet it is not merely by thefe pictures that
we get the idea of the fhapes of thofe objects;
the eye only confiders a point of an object at
a time ; and it is by the motion of the eye itfelf, or of the body, tracing its boundaries, that
we get the idea of its fhape ; fo that it is done
by feeling, not by colour merely, as others
have already obferved. Every one who has
been weakened by a fever will remember how
painfully the fenfe of feeling is excited by
ftrong founds, and the particular manner in
which they affect the head.   In general, therefore, if an aërial found comes from a certain
quarter without us, it feems to be made, by
the contrivance of the external ear, to beat
upon a particular part of the tympanum ; the
found, anfwerable to the time of vibration, is
excited in the labyrinth ; and at the fame time,
a like tremulous fenfation of feeling as in that
                                        part

part of the head which anfwers to the fituation of the external found, and this, from cuftom, gives us the idea of the found coming from
fuch a quarter. And if two or more founds
come from different quarters without us at the
fame time, they feem to be made, by the funnel
of the ear, to beat upon different parts of the
tympanum; and befides the founds which they
excite in the labyrinth, caufe refpectively unifon or like tremulous fenfations of feeling, as
in the anfwerable fides of the head; whence
we come to underftand their fituations externally with refpect to each other and to our-
felves. The tympanum may move as a whole,
notwithftanding thefe vibrations excited in
different parts of it ; and (by the mediation of
the little bones) *according as its motion as a
whole is affected by thefe particular vibrations,
the paffions or affections of the mind and body
may be influenced.* I do not know whether the
harmony and difcord of founds, confidered mu-
fically, may not partly depend on this latter
principle.

Obs. IX. That it is by the fenfe of feeling
as above, or, at leaft, that it is not by mere

found that we get the ideas of the fituations of
external founds, feems alfo to appear by the
following confiderations. Admitting that different fides of the tympanum are affected, as
above, according as founds come from different
quarters, yet the nature of aërial vibrations,
and the fituation of the tympanum with refpect to the paffage into the labyrinth, feem to
be fuch that the founds cannot be directed in
right lines from the different parts of the tympanum affected, through that paffage, as the
rays of light crofs in the eye and pafs on to
the refpective parts of the retina, for painting
objects anfwerable to their fituations out-
wardly; and even fuppofing they could, yet
the ftructures or fhapes of the cochlea and labyrinth are fuch as that no advantage could
be derived from this rectilinear direction;
they feem to enter into the organs in a confufed medley, through every part of their wind-
ings, and alike from whatever quarter they
come; and it is only the internal founds, de-
pofited in thofe organs, that feperate them,
according as they happen to be in unifon.
Sound, therefore, feems to have no other con-
cern in the affair of hearing than merely as to

the

the note or tone, and the ſtrength or loudneſs
thereof, as obſerved before.

OB s. X. Diſtinct hearing ſeems to be concerned only with a particular part of the tympanum, as diſtinct viſion is only with the middle
of the eye. If we would view an object diſtinctly we turn the optic axis towards it, and
even to the ſeveral parts of it ſucceſſively; and
if we would hear any ſound to perfection, we
turn our face towards the quarter from whence
the ſound comes; in ſuch poſition of the face,
the ſound falls perhaps on the middle of the
tympanum; or at leaſt on that part of it which
ſerves for diſtinct hearing. Do the ſounds in
this caſe fall upon the malleus?—or rather, do
they not fall in the midway between the malleus, and the margin of the membrane? At
leaſt, do they not beat upon ſuch a part of the
tympanum, as that they may be thrown to moſt
advantage into the feneſtra ovalis, or elſe upon
the elaſtic membrane in the rotunda?

OB s. XI. Diſtance, with regard to hearing,
ſeems to depend on much the ſame principles
as diſtance with regard to viſion, viz. on the

faintneſs

faintnefs and indiftinctnefs of the found beating on the tympanum, &c.

Obs. XII. Though we have two ears, yet
we hear fingly, even as we fee fingly with
both eyes. Are the nerves for the fenfe of
feeling, which ferve the tympanum (or that
part of the ear, whatever it be, which gives
us the idea of the fituation of founds), joined
together in the brain, as the fibres of the retinæ of the anfwerable parts of both eyes are
fuppofed to be? or is each ear concerned only
with the anfwerable fide of the fenfory? or do
we hear only with one ear at a time, as fome
have fuppofed that we fee only with one eye
at a time?

Obs. XIII. May it not be on account of the
diftance of the labyrinth and cochlea from the
outward ear, that the innate founds are diffi-
cultly excitable by external preffure, and that
many of them cannot be excited thereby at all?
the deepeft of thefe founds may be in the cochlea, and the higheft in the labyrinth; for the
high and low founds feem to be in different
parts of the ear. As I cannot excite founds
lower

lower than G, is it not that only the higher
ones in the labyrinth, and not the lower ones
in the cochlea, are excitable by outward preffure ?

OBS. XIV. Thofe who are difficult of hearing, and alfo fuch as liften attentively, are apt
to open their mouths, and then they hear better. It has been faid that in this cafe the found
paffes by the Euftachian tubes ; but whoever
choofes to make the experiment will find, by
putting a finger into each ear, that when he
opens his mouth, the bones of the lower jaw
leave the meatus auditorius much wider than
when the mouth is fhut, which may perhaps
be the caufe of the phenomenon, at leaft in
part.

OBS. XV. The vibrations in the ear by
which the internal founds are caufed, are not
of the nature of thofe of mufical ftrings, for
the following reafons : the length of the fibres,
efpecially for the low notes, would be too
great for them to be contained in the ear, if
they caufed founds by vibrations in the manner
of mufical ftrings. If it be objeted, that their

C 4   tenfion

tenfion is proportionally lefs, I anfwer, that then by reafon of their flacknefs they could not vibrate, on that principle, at all. *2dly*, There are no fibres in the ear but what are immediately furrounded by fuch fubftances as would totally hinder a vibration in the manner of mufical ftrings. *3dly*, There is no trace of any contrivance in the ear, which can in the leaft favour a fuppofiticn of this kind to him who properly confiders the above objedions ; and there are many others which may be urged, but thefe are I imagine fufficient. In what particular manner thefe vibrations are performed I cannot determine: it may be by means of fibres whofe particles or elements, when irritated, alternately approach towards, and recede from, each other, and thereby lengthen and fhorten the fibre by turns, without forming the harmonic curve like a mufical ftring. The fibres for the deeper notes may be compofed of larger particles, and thence vibrate more flowly ; and on the contrary, the fibres for the higher notes may be compofed of fmaller particles, and thence vibrate more fwiftly ; the pulfes in both cafes being communicated to the nervous fibrils with which they

are

are refpectively ferved: the like may be the
cafe in the eye, and the fenfe of feeling. A fingle
ftring, or even two particles only, feem proper
for this purpofe, and they would alfo take up
lefs room in the organs; and be too minute
perhaps to be difcovered by the eye. Suppofing this to be true, we have a reafon why in
the experiment of prefing the centre of the
eye, the green, blue, and violet-making vibrations are moft eafily excited, and that their
nerves are laft paralytic; and contrariwife
with the yellow, orange, and red-making ones.
The particles of the former being lefs, are more
eafily put into vibrations, and lefs apt to prefs
on the nervous fibrils than thofe of the latter, which are larger *; and this may alfo be
one reafon why the low innate founds cannot

---

* By prefing the centre of the eye for fome time, a luminous ring of a reddifh yellow colour is often perceived; if
the prefure be removed, and the eye turned towards the
light of the clouds, the ring (which does not difappear for
fome time) is changed into directly oppofite colours, viz.
green, blue, and violet; which fhews that the prefure had
rendered the former internal colours more paralytic than the
latter, agreeable to what was faid before. M. Le Cat, if I
remember right, has experiments to the fame purpofe, which
may therefore be explained by this theory.

be

be excited by outward preffure, the particles
of the fibres, being large, requiring a ftronger
irritation than can be there applied. But this
is conjecture.

Obs. XVI. There are (it may be prefumed)
many octaves of the internal founds ; but not
quite one octave of colours. This difference
was requifite, becaufe there may be a great
number of vibrations made in the air, which
would be loft to us, if there were not anfwerable ones in the ear ; whereas the vibrations
of the rays of light being limited to about a
fixth, only the like latitude of internal vibrations was required anfwerable to them.

Obs. XVII. In the eye there feem to be
a great number of vibrations which give the
fame colour difperfed in every part of the retina ; and vibrations, in all the different times,
feem likewife to be mixed equally together in
all parts of that organ. Thus, if any part of
the retina be excited by preffure, not a fingle
colour arifeth (unlefs by accident, as in the inftances related in the firft fection,) but a white
one, compofed of all the others. But by exciting

ting the innate founds I can hear many of them
diftinct ; nor can a found be excited compofed
of all the internal ones, as is the cafe with colours. The reafon of this difference is, that
the internal colours can be mixed together in
the retina, in a fpace fmall enough for them to
be perceived only as one colour. But this cannot take place with regard to the internal
founds, by reafon of their far greater number.

Obs. XVIII. About ten years ago, I obferved that a flute, an hautboy, a trumpet, and
other inftruments, though they were made to
yield founds which were in unifon with each
other, and equally loud, yet had a difference
which every one could obferve, and which I
then called the *mode* of found. Thus alfo the
voices of people, and the founds yielded by various bodies, though of exactly the fame tone
and ftrength, had a fimilar difference. Whether the caufe of this curious phenomenon had
been difcovered, I could not learn ; but by
meditating on the fubject, and making feveral
experiments, I found that thefe founds were not
fimple, but compofed of others, of which thefe
were only the refult or aggregate, even as the
                                        colours

colours of bodies are various compounds of the feveral original colours. I am told, by a gentleman to whom I communicated this theory, that a difcovery of this kind is already made public, though I have not yet been able to get a fight of it : I fhall not give the obfervations which fuggefted that theory to me, left they fhould be fimilar to thofe alluded to *. But as I cannot find that an explanation has been

* The principle on which thefe founds were capable of being decompofed was, that in many cafes fome of the founds in the mixture were ftronger, and others weaker: hence if they were excited as gently as poffible, or rather if I removed the caufe of them to a fufficient diftance from me, the weak founds in the compound would be inaudible, and only the ftronger ones heard. Suppofe a found compofed of C D and E, that C was double the loudnefs of D, and D double that of E ; the tone of that found would be in a compound ratio of the tones and ftrengths of the ingredients ; that is, it would be a fharp C. If E be rendered inaudible, the found, as being compofed only of C and D, would be nearer to C ; but if D alfo be made inaudible, the found would be pure C. If the ingredients are equal in ftrength or loudnefs, this decompofition cannot be made. This theory was fuggefted to me by like obfervations with regard to colours ; for fome objects, according as they are more or lefs ftrongly illuminated, appear differently coloured, for reafons (as I imagined) fimilar to thofe given above.

given

given of this phenomenon with regard to what
happens in the ear, I fhall here fubjoin a conjecture concerning it : it was fhewn above,
that all the internal colours may be excited
within a fpace of the retina fmall enough for
them to be perceived only as one colour, the
aggregate of the whole. If aërial founds are
blended together, as in the cafes juft mentioned, they are excited fo near to each other
in the tympanum, that they are heard only as
one found, as happens with a mixture of colours; but if the founds are excited at a diftance from each other in the air and tympanum, they are heard diftinct, as happens when,
in vifion, colours are painted in different
parts of the retina. But to enter a little clofer
into this reafoning, thofe who are verfed in
optics know, that if any two neighbouring colours in the refracted fpectrum be mixed together, the colour arifing therefrom will be fuch
an one as would be caufed by the rays in the
mean betwixt them : thus, if blue and yellow
be mixed together in equal quantities, the colour will be green ; and if the quantities be
unequal, the green will be tinctured with yellow or blue in proportion; the like may be
obferved

obferved of other neighbouring colours. But if red and violet be mixed together, the colour will not be a green, or the intermediate one, but a kind of purple, unlike to any of the original colours. Alfo, if any number of colours are mixed together, provided the two extreme ones are at a fufficient diftance from each other in the fpectrum, there will not be produced the intermediate prifmatic colour, but fome one unlike to any of thefe : thus, a mixture of all the rays compofe a white, and fo of other mixtures; for further information in which, Sir Ifaac Newton's Optics may be confulted. Now, I would fuppofe that a fingle feries of colorific vibrations in the retina are difpofed in a right line according to their times, as in a refracted beam of light, and that this line exceeds the diametre of a vifible point, yet is not fo long as that the two ends of it may be perceived diftinct. Hence the red and violet only, though they are not feparately diftinguifhable, yet as they do not fall within a vifible point, they alfo cannot be perceived as a perfect mixture or under the form of the intermediate colour; they muft therefore be perceived as in a ftate between perfect mixture, and diftinctnefs ; and

we

we find that a purple is the refult, a colour in
which the ingredients can in fome meafure be
inferred by the eye. But two colours which
are near each other, are contained within a
vifible point, and therefore may be faid to be
mixed intimately together, for they exhibit
the proper intermediate colour, as was fhewn
above. For the fame reafons, all the colours
in the feries together ought not to exhibit,
like yellow and blue, the intermediate colour,
nor any of the original ones, becaufe of the
red, violet, &c. which are exterior to the vifible point; neither ought the colour exhibited to be fuch an one as that the ingredients
may in a manner be inferred by the eye, as is
the cafe with red and violet alone, becaufe the
whole feries is a compofition of perfect and
imperfect mixture; and we find that they
compofe a white * Now if we apply thefe principles

---

* If we fuppofe a number of thefe feries joined together
in a right line by their anfwerable ends, viz. red to red, and
violet to violet, that the whole furface of the retina is filled
with fuch lines drawn parallel to each other, and that thefe
lines are croffed, at the red and violet points, at right angles
by fimilar ones refpectively; the whole furface will be divided

ciples to the ear, and confider hearing as compofed of found, and a tremulous fenfe of feeling, as mentioned before, we may be enabled
to form fome idea of the caufe of the above-
remarked difference of founds, whofe tones and
ftrengths are the fame. The innate founds
are not, like colours, comprized within fo fmall
a portion of the ear as that they may be all
heard as one: on the contrary, experiment
feems to fhew that they are *diftinct*. The above
doctrine, therefore, is not applicable to the innate founds; it muft of courfe be applied to
that part of hearing which depends on the tremulous fenfe of *feeling*, and by which the other
is governed, as hath been fhewn. If two or
more aërial vibrations fall within the fame
point of the tympanum, they may be confidered
as mixed perfectly together *; and therefore
a found

ded into fquares, the fides of which will be lefs than the
diftances required for diftinct vifion.———But this is a mere
hypothefis.

* Query. Whether in any inftance of this kind, two or
more vibrations are converted, as it were, into one?—For example, whether a found and its octave make, not the intermediate found, but the octave? I do not think this ever to
be the cafe; but that they continue diftinct, and therefore
the

a found will be caufed whofe note is the mean
of all thefe, and whofe mode is the fame with
that which would be produced by a fingle vibration. But founds in the tympanum further
apart, yet not fo diftant as to be heard diftinct,
though they yield a found of the intermediate
tone, yet the mode thereof, by reafon of the
imperfect mixture, fhall be different. And if
the founds are excited at a ftill greater diftance,
they fhall be heard diftinct; and therefore by
affuming the hypothefis of the tremulous fenfe
of feeling, and carrying with us the idea of
perfect mixture, of indiftinction or imperfect
mixture, and of diftinct founds, as above, duly
combining thefe, varying their ftrengths, and
taking into confideration what happens in the
cochlea and labyrinth with regard to the innate founds, and alfo the paffions or affections of
the mind, we may have perhaps a theory of this
kind of phenomena. But as this is a fubject
which does not eafily admit of experimental
proof I fhall not enlarge on it.

the intermediate tone is the refult. The reafon why fome
modes are more pleafing than others, may perhaps be collected from the eighth Obfervation.

D   S E C T.

SECTION V.

Being an Appendix to the foregoing Effay.

IT may not be improper to acquaint the Reader, that my fituation in life has hitherto been fuch as to have afforded me but few opportunities of reading books which were not in the line of my profeffion ; many difcoveries in philofophy therefore had been made public which I was unacquainted with : many I fuppofe ftill remain, of which I have no idea. As I am fond of amufing myfelf at my leifure with ftudies of a philofophical nature ; ideas have occurred to me which I thought were new, but which I have afterwards found in authors. The doctrine of the modes of found occurred to me above ten years ago, (as indeed did almoft the whole of the preceding effay). I mentioned that I had fince fhewn it to a friend, who informed me that the difcovery had already been publifhed by Tartini, an Italian.

Italian. But fince that effay was fent to the prefs, I happened to meet with a tranflation of Roufleau's Mufical Dictionary, and find by it that the theory in queftion is not yet known, Tartini's difcovery being of another kind, viz. *The harmonic founds which arife in confequence of any mufical found;* and on which he has founded a new fyftem of mufic.

M. ROUSSEAU, after explaining the two differences of founds, the *tone* or *note*, and the *ftrength* or *loudnefs*, fpeaks of this third difference of founds; and expreffeth himfelf as follows:

" IN regard to the difference which is found alfo between the founds by the quality of the modification, it is evident that it holds neither to the degree of elevation, nor even to that of the force. It will be in vain for an hautboy to place itfelf in unifon with a flute : it will be in vain to fweeten the found to the fame degree; the found of the flute will always have a *Je ne fais quoi* of mellow and fweet ; that of the hautboy fomewhat rude and fharp, which will prevent the ear from confounding

D 2 them,

them, without mentioning the diverſity of the
modification of the voice.

" Th e r e is not an inſtrument which has
not its particular tone which has no connection
with that of another; and the organ alone has
twenty methods of playing, all of a different
modification. *No one however that I know of
has examined the found in this particular,* which,
as well as the reſt, will perhaps be found to
have ſome difficulties; for the quality of the
modification cannot depend either from the
number of vibrations which forms the degree
from flat to ſharp, or from the greatneſs or
force of theſe ſame vibrations which forms the
degree from ſtrong to weak: we muſt then
find in the ſonorous body, a third different
cauſe from theſe two, to explain this third quality of ſound and its differences, which perhaps
is not too eaſy." Thus far Mr. Rouſſeau.

I s h a l l therefore give the Reader the ob-
ſervations from which I was afterwards led to
ſuſpect this theory, and in the very words of
the original paper which I have long had by
me.

" ——Th e

" ————————The innate founds are the moft
fimple poffible, as they are made (we may fuppofe) only by fingle or fimple vibrations;
whereas the aërial founds which we hear, are
all of them more or lefs compounded. And on
this compofition of them, the folution of fome
curious phenomena feem to depend, which have
not been attended to by the learned.

"For befides that founds are higher or lower
in tone according to the fwiftnefs of the vibrations which caufe them, there is alfo another
property in which founds differ from one another, though of the fame ftrength and tone :
an organ, for example, has feveral kinds of
ftops; there is one which being played re-
fembles a flute ; another which gives the fame
mufical notes as the above, yet differ very much
in found : for whereas thofe refembled a flute,
thefe found like a trumpet ; another ftop gives
the fame mufical notes as both thefe, yet re-
femble an hautboy. The like may be inftanced
in bells, the ftring of an harpfichord, violin,
&c. So I can with my mouth found a row of
notes, or fing a fong in a manner refembling a
flute ; I can found the fame notes like a trum-

pet,

pet, an hautboy, and other inftruments: fo in converfation, I can talk in a great many different voices, though with the fame tones and loudnefs. There are hardly two people in the world whofe voices are exactly alike, though they were to talk in the fame mufical tones, and equally loud; and it is from hence that we know the voices of people who are talking to us. The founds alfo which are yielded by bodies that are ftruck are different, though of the fame tones and ftrength, infomuch that there are hardly two kinds of bodies which found exactly alike. This difference of founds I call the *mode* of found to diftinguifh it from the *tone* and *ftrength*.

" OBSERVATION I. If I ftand by a large church bell when it is ftruck, and liften attentively to the found when it is almoft vanifhed, I can diftinguifh not one found or note only, but feveral; for the real note of the bell will go off and be no longer heard, and inftead thereof other founds different in *tone*, as well in *mode*, will arife, which in fome bells are more and in others lefs in number: and different

bells

bells exhibit this phenomenon with different degrees of diſtinctneſs.

" Obs. II. I lived near a church in which were eight bells, and the clock ſtruck on the the firſt four of them every quarter of an hour, I have frequently obſerved that if I was ſo ſituated as that the ſounds could hardly be heard, or heard indiſtinctly, the ſounds of theſe bells, which otherwiſe were D, C ſharp, B, A, in theſe caſes totally loſt their ſucceſſion; they gave quite different tones, as G, E, C, B, or others, and their *modes* likewiſe were different.

" Obs. III. I have obſerved the like in the voices of people under ſimilar circumſtances, and alſo in the ſounds produced by various other methods. I mean in caſes where the ſounds were hardly ſenſible, or indiſtinctly heard.

" Obs. IV. I once ſaw a peal of ſmall houſe bells, with which a young gentleman uſed to amuſe himſelf; if I ſtood near theſe bells when he rang them they were very tunable, and made good muſic; but if I removed to a diſtance from them, though I heard the bells

D 4   diſtinctly

diftinctly ftruck, they no longer yielded founds
in fucceffion, every one a note lower than the
preceding, but quite irregular and confufed :
the irregularity was different at different diftances, and the *modes* were altered as well as
the tones.

" FROM thefe obfervations I gather *that the
tones of founds which are yielded by bells, &c.
are not fimple, but compofed of other tones.* Thus
in optics, the rays of light in the fourth feries
of a fpectrum caufe a green colour, which
being produced by one fort of rays only, may
be termed fimple ; the rays of the third feries
are yellow, and of the fifth blue : yet when
mixed equally together, they no longer appear
as yellow and blue ; but a colour refults from
their mixture, which is the fame as would be
caufed by the fimple rays of the fourth feries.
If thefe rays are again feparated, they no longer
caufe a green, but their proper colours ; if
the rays are mixed in unequal quantities they
caufe a green, not like the other, but inclining
to the colour of the greateft quantity of rays.
Alfo, if either in equal or unequal quantities
they are mixed not perfectly together (fo that

for

for example, if the object from which they come be viewed near, or but faintly illuminated, thefe colours may be feen diftinct; yet if viewed at a diftance, or the object be ftrongly illuminated) they caufe an imperfect green, fo all the feven original colours mixed together produce a white, and the like.

" Bodies in general do not appear exactly of the fame colours when viewed by a ftrong, as when viewed by a faint light, which I take to arife from hence; that bodies do not reflect an equal quantity of each fort of rays. Suppofe that a body reflected four parts of red, three of orange, two of green, and one blue; if the light be fo faint as that the red rays chiefly be fenfible, the object will be redder; and if the light be ftill ftronger, the colour will vary from that red with the increafe of the illumination, till it appears of that colour which ought to refult from the above mixture,

" For the fame reafons, if two founds are mixed equally together, and of equal ftrengths, but of different tones, they will caufe a found whofe tone is the mean between the tones of the two founds when feparate : thus G and B

being

being mixed, the note will be A; if they are
unequal in ftrength, the tone will incline towards G, or B in proportion ; and all the other
inftances of colours above given may be applied
in fome meafure to founds.

" So then the reafon of the change of tones
of found by diftance, inattention, or the like,
arifeth from hence ; that if the founds which
are blended together be of unequal ftrengths,
thofe which are ftrongeft muft reach to a greater diftance than thofe which are weaker ; fo
that the weaker ones not affecting the ear, or
not with fufficient force, only the ftronger
found or founds which reach the ear will be
perceived; but if the founds are equally ftrong,
this will not take place.   The founds of the
bell in the firft obfervation, however, are to be
underftood by what was faid above of colours
being mixed imperfectly together, and therefore they were not heard as one found but
when they were fufficiently ftrong, fo as to
fpread their effects on the fenfe into one another *.

" I<small>T</small>

---

* The note of latitude, as it is called, in wind inftruments,
may depend on the principle of the mixture of founds of different

" It has ufually been thought that all founds affect the fame parts of the ear; but the fact appears to be otherwife. And it is furprizing how near founds may be to each other in the air, and yet be heard diftinct, and even when no longer heard diftinct, they are preferved feparate: it does not appear that two founds form only one vibration when thus mixed, but the intermediate tone. A theory of aërial founds in this view therefore is yet wanting, as well as that of different founds yielded at the fame time by the fame body.

" If the founds thus mixed together are concords, they form perhaps the fweeteft modes; and on the contrary, the difagreeable modes feem to be compofed of difcords. The laft five or fix of Bow bells are, I think, the moft agreeable in peal of any that I ever heard; and the reafon is that the founder has judicioufly varied the *modes* as well as the *tones*. This obfervation might probably be applied to good ufe by lovers of mufic.

Now thefe complex founds, though we per-

ferent ftrengths, all of which are not audible but by a ftrong blaft.

ceive

ceive them as one, do not, I take it, excite that innate found in the ear which anfwers to the tone we hear; but every one of the vibrations which compofe that complex found excites its unifon, which are fo mixed together in the fenfe that we perceive only one found, the refult of the whole, as before obferved.———"

WHAT relates to the fenfe of feeling (of which I had no idea at the time of writing the above) has already been explained, and which feems to have a great fhare in thefe modes, infomuch that the mixture of the founds, and their being heard as one, depends on it, and not on the innate founds, as hath been fhewn. The innate founds may be excited in greater or lefs number; ftronger or weaker; and in more or lefs harmonic or difcordant relation to each other, which feems to be all that found contributes to thefe modes. The tremulous fenfe of feeling excited in the tympanum, the parts connected with the little bones, the portio dura, the nerve that adheres to the tympanum, the various diftributions of the twigs of the nerves of the ear, fo as to form fympathies with other parts, and the affociation of ideas

are,

are, perhaps, all concerned in thefe modes, though chiefly the firft. I have only begun the fubject, and would wifh to fee it further profecuted by thofe who have leifure and inclination. In the mean time it may, I think, be admitted, that " as the colours of bodies are not fimple, but made up of others, according to the different mixtures of the rays of light iffuing from them, fo neither are their founds fimple, but compofed of feveral or many others, which the body by its various vibration emits; and which, like the colours, are fo mixed together in the fenfe, as to appear but one, the mean of all the ingredients. The *modes* of the founds depend on the manner of this mixture."

A N

A

TREATISE

O N

HARMONIC SOUNDS.

INTRODUCTION.

AS I have had occafion to mention Mufic in feveral places of the preceding Effay, the following may be inferted with fome degree of propriety.

Some time ago I found out the theory of the harmonic founds yielded by mufical ftrings, and conftructed a fcale of them for four ftringed inftruments tuned fifths, without then knowing that they had been publifhed long before. I have fince been undeceived by the friend who informed me that Tartini had difcovered the theory of the modes, and by Mr. Rouffeau's Dictionary. There are two very material points, however, with refpect to the practice of thefe founds, which I cannot find any account of in that Dictionary; and therefore, as my paper on this fubject is fhort, I will fubjoin the whole of it.

THE

THE

THEORY

OF

HARMONIC SOUNDS.

THE *common* found of a mufical ftring
is caufed by a fimple or fingle vibration
thereof; the *harmonic* found by a various vibration.

IF one of the ftrings of a violin be ftruck
with the bow open, it vibrates, and thereby
yields a found : if the finger prefs the ftring in
the middle upon the finger board, its lower part
only will vibrate, and its vibration will be
twice as fwift as that of the whole ftring, fo
that the found will be an octave above the
former.

BUT if the finger be laid lightly on the
ftring, without preffing it on the finger board,

both

both halves of it will vibrate ; their vibrations
will coincide with each other ; the found arifing therefrom will be much fweeter than in
the other cafe, and will be the harmonic octave
to the open ftring

If the finger be placed at one third of the
ftring from the nut, and ftruck with the bow,
it will vibrate in three diftinct and equal parts,
coincident with each other. You may prove
this by laying another finger lightly at two
thirds ; for though the ftring be thus ftopt
double, the found will be the fame; though if
you remove either or both of the fingers from
thefe points, either higher or lower on the
ftring, the note ceafeth. Alfo, if with a finger
at $\frac{1}{3}$, you bow at $\frac{2}{3}$, no fuch found will be excited ; but if you remove the bow fufficiently
from this point either way, the found again a-
rifeth *. The vibration in this cafe being to
that

* You may fee this threefold vibration, at leaft in the filver
ftring: but in a bafs viol you may fee it much plainer, and
there the four, five, and fixfold vibrations, may alfo be dif-
tinguifhed by the eye. Each harmonic ftop is not confined
to the point, but has a latitude, and the points of the $\frac{1}{2}$ and $\frac{1}{3}$
divifions have a greater latitude than the leffer divifions. The

E   reafon

that made by ftopping at one half, as 3 to 2 makes a fifth above it.

PLACE the finger at one fourth of the ftring from the nut, it will vibrate in four diftinct and equal portions, coincident with each other, and being twice as fwift as the firft, the found will be an octave above it.

REMOVE the finger fucceffively to $\frac{1}{5}, \frac{1}{6}, \frac{1}{7}, \frac{1}{8}, \frac{1}{9}$, of the ftring, it will vibrate in 5, 6, 7, 8, and 9 equal parts ; and you will, for reafons fimilar to thofe given above, have a fharp third, a fifth, and very fharp fixth (or flat feventh) to the octave; an octave to that, and a full tone above that double octave, and you may go ftill higher, by taking the $\frac{1}{10}, \frac{1}{11}, \frac{1}{12}$, &c. of the ftring.

THE above notes may be made by ftopping at any other point befides $\frac{1}{3}, \frac{1}{4}, \frac{1}{5}$, * &c. and like-

reafon of this latitude is, that near the points the motion of the ftring is very fmall, and therefore is not much interrupted by the finger ; whereas, nearer the middle, the vibration is more eafily ftopt. The tone is alfo a little altered when the finger is not exactly on the point; the reafon is obvious.

* Any of thefe frets may be ufed in practice, when more convenient than the other.

wife

wife by ftopping at all, or more than one of thefe divifions: thus, the fifth above the key may be made by ftopping at $\frac{2}{3}$ from the nut, as well as at $\frac{1}{3}$, and alfo by ftopping at both $\frac{1}{3}$ and $\frac{2}{3}$. The octave may be made by ftopping at $\frac{1}{2}$ or $\frac{3}{4}$ as well as at $\frac{1}{4}$ from the nut ; and alfo by ftopping at all, or more than one of thefe points (which may be done by threads faftened acrofs the ftring round the inftrument). The fharp third above the octave may be made by ftopping at $\frac{2}{5}$, $\frac{3}{5}$, $\frac{4}{5}$, as well as at $\frac{1}{5}$, and likewife by ftopping at all, or more than one of thefe points ; and fo of the reft. Whence you have this caution; " that if you do not happen to ftop right, fome other note than that intended may arife." Thus, if you place your finger a little below $\frac{2}{3}$ from the nut, you light on one of the $\frac{3}{5}$ divifions, and fo have, inftead of a fifth, the fharp tenth: a little lower you fall on one of the $\frac{4}{7}$ divifions, which gives the fharp 13 (or flat 14) and fo of others; of which therefore you muft be aware,

ALSO you muft be careful not to bow upon the points or divifions of the ftrings, for then either no found will arife, or not that defigned,

but

but between thefe points: thus, if you ftop at
$\frac{1}{4}$ you muft not bow at $\frac{1}{4}$, but between that and
the bridge, or between other points, though
that next the bridge is beft, the ftring being
moft fteady there. From whence alfo it appears " that you muft bow nearer to the bridge
in proportion as you ufe an higher fret," the
divifion being lefs.

LIKEWISE when you ftop at $\frac{1}{2}$, you will,
inftead of the key, often get its octave, unlefs
you bow towards the verge of $\frac{3}{4}$ from the nut,
becaufe the $\frac{1}{2}$ fret is alfo a $\frac{1}{4}$ divifion *   But by
bowing near to $\frac{3}{4}$ you do not excite the octave,
for reafons which may be feen above; and fo
of other notes.

A STRING is fo apt to run into harmonic
vibrations, that thefe founds may be raifed
merely by bowing on proper parts of it, without ftopping with the finger: thus, if you
bow on the proper parts of the filver ftring
near the bridge, you have thirds, fifths, eighths,

---

* It is likewife a $\frac{1}{6}$, $\frac{1}{8}$, $\frac{1}{10}$, &c. divifion. The like may be
obferved of the $\frac{1}{3}$, and other frets: and the various refpective
founds may be raifed by bowing properly, as above.

and

and other harmonic notes; and they may likewife be raifed by bowing on other parts of the
ftring, by obferving what was faid in the preceding paragraph. The bow, in thefe cafes,
acts in a double capacity, for it both ftops and
vibrates the ftring *.

FROM what has been faid, it appears " that
the harmonic founds are made by ftopping the
ftring lightly, according to the proportions in

* I find by the Dictionary of Mufic, that Tartini has
founded his fyftem on this obfervat'on : I am miftaken,
however, if he has not proceeded on a wrong principle. He
fays (if I remember right, for I have not the book now by
me), that when a ftring is founded, all the notes harmonic
to the found naturally arife with it; and he applies it to all
other founds. It is true that if you ftrike a ftring with a
bow, you will often raife fome of the harmonic founds, for
a reafon given in the laft·paragraph ; and that a ftring fhould
vibrate as a whole, and in diftinct parts at the fame time, is
as eafy to conceive as that the moon can revolve at once
round the fun and our earth. If, however, you excite the
found of a ftring by any other means, as by ftriking it with
a ftick, or pulling it with the finger, I do not find that any
fuch founds arife. I have not yet had leifure to fatisfy myfelf concerning this matter; but mean to examine both this,
and what Tartini fays of the *third founds*, when I have a convenient opportunity.

E 3   the

the following feries, viz. $\frac{1}{2}, \frac{1}{3}, \frac{1}{4}, \frac{1}{5}, \frac{1}{6}, \frac{1}{7}, \frac{1}{8}, \frac{1}{9}$, &c. the ftring vibrating in as many diftinct and equal portions as the denominator hath units, all in unifon with each other; and the founds being higher according as the portions of the ftring become fhorter; that is, according to the fwiftnefs with which thofe parts vibrate." This may fuffice for the theory of thefe founds, we may now proceed to

THE PRACTICE.

THE ftrings of the violin, &c. being tuned fifths *, the harmonic notes on them will be as in the following fcale.

* By means of thefe founds the inftruments may be tuned to the greateft exactnefs : to do which you have only to fcrew up the ftrings fo as to bring the 5, 9, 13 on the line $\frac{1}{2}$ in unifon with the 5, 9, 13 in the line $\frac{1}{3}$ (fee the fcale); and as the ear can better judge of an unifon than a fifth, you may tune to greater perfection than in the common way. This alfo I could not find in Mr. Rouffeau's Dictionary.

SCALE.

SCALE.

Fourth String	1	5	8	10 Sh.	12	14 Fl.	15 &c.
Third String	5	9	12	14 Sh.	16	18 Fl.	19 &c.
Second String	9	13	16	18	20	22 Fl.	23 &c.
First String	13	17	20	22	24	26 Fl.	27 &c.
The divisions from the nut	$\frac{1}{2}$	$\frac{1}{3}$	$\frac{1}{4}$	$\frac{1}{5}$	$\frac{1}{6}$	$\frac{1}{7}$	$\frac{1}{8}$ &c.

You may carry it still higher, by adding frets above $\frac{1}{8}$; but this commands a sufficient compass of notes for practice. It has, however, the inconvenience of being incomplete, especially in the lowest and best notes.

WITH a view therefore to improve this scale, or obtain one more perfect, imagine the strings to become continually shorter, or that the bridge and nut approach toward each other with a regular motion; the divisions of $\frac{1}{2}, \frac{1}{3}, \frac{1}{4}, \frac{1}{5}, \frac{1}{6}$, &c. will still remain in the same proportions, but on a scale continually contracting or lessening; and the sounds will become higher in proportion to the shortness of the strings.

E 4   To

To reduce this to practice, place the little finger lighty on $\frac{1}{3}, \frac{1}{4}, \frac{1}{5}$, or any other harmonic fret of either of the ſtrings; place the fore finger on the nut, and ſtrike the harmonic ſound with the bow, continue the bowing while you ſlide your fore finger from off the nut upon the ſtring, preſſing it down hard on the finger-board, and from thence along the ſtring up towards the bridge, bringing your little finger nearer towards it, ſo as that you may be always at $\frac{1}{3}$ (if you uſe that fret) of that part of the ſtring between the fore finger and bridge, ſo ſhall you have a continually aſcending harmonic ſound. From whence it appears, "that you may make an harmonic ſound of what degree of ſharpneſs or flatneſs you pleaſe" (within the compaſs of the inſtrument) : with ſounds made after this manner, therefore, you may ſupply the deficiencies of the above ſcale, at leaſt from 5, upwards; whereby you may make it as perfect as you pleaſe.

Or you may compoſe your ſcale intirely on this plan, (though it muſt be owned that the notes are leſs harmonious than when the ſtrings are not hard ſtopt) ; thus, make G, A, B, C, on

the

the filver ftring as in the common way, ftopping hard for thofe notes with the fore finger,
and making them harmonic by ftopping lightly
with the little finger at ⅓ of that part of the
ftring between the fore finger and bridge, and
fo on with the other ftrings. By this means
you have a compafs of fixteen notes, ufing only
one fret, and going no higher than A on the
treble ftring. But by fhifting the hard ftop to
B, C, D, &c. you may go ftill higher; and
higher after all by changing the fret for thofe
above. You may alfo fhift on the other ftrings,
and on any part of any ftring you may by this
means make not one only, but as many harmonic founds as your fingers can command frets.
The practice indeed is fomewhat difficult, but
can be done, I imagine, fufficiently well by one
ufed to fhifting and double ftops; or, in other
words, by a mafter of the violin.

BUT the beft fcale for practice that has yet
occurred to me is the following.

A SCALE

A SCALE of the harmonic notes of the violin,
according to the Diatonic Genus ; which
therefore might eafily be varied for the other
genera, and alfo for other inftruments of the
the viol kind.

Take only the following notes of the former
Scale.

G	D	A	E	⊢	ω
*	*	*	*	⊢	ᴎ
*	*	*	B	⊢	o
B	Fˢʰ·	Cˢʰ·	Gˢʰ·	⊢	ᴧ
G	D	A	E	⊢	ᴛ
D	A	E	B	⊢	ᴍ
G	*	*	*	⊢	ᴎ

Thefe notes run thus : G, *, *, *, D, *, *, G,
A, B, *, D, E, F fharp, G, A, B, C fharp, D, E, *,
G, A, B, *, *, E.

I HAVE rejected the D, A, E in the line ½,
and alfo thofe in the line ⅙, becaufe they are
more convenient for playing in the lines ⅓ and
¼ : I reject all the notes above ⅛, and likewife
thofe

thofe in $\frac{1}{7}$, becaufe they are too difficult to hit, and becaufe thofe made by the hard ftop are more harmonious. Perhaps thofe on $\frac{1}{8}$ might alfo be rejected for the fame reafon. The vacancies may be filled up as follows.

THE fingers not being long enough to complete the notes from G to D, the fcale can only be perfected from D upwards.

IN order to this, place your fore finger on the filver ftring, as for making the common G fharp, preffing the ftring down on the finger-board, as in the common way, at the fame time lay your little finger as lightly as poffible on the ftring at one third part of the diftance between your fore finger and the bridge, found with the bow, and you have the harmonic D fharp.

REMOVE the fore finger, as for making the common A, and the little finger a little farther on, you have E.

MAKE common A fharp with your fore finger, and remove your little finger fomewhat nearer to the bridge, you have F

MAKE

MAKE the common B with your fore finger, and remove your little finger a little farther on, you have F fharp.

MAKE the common fharp C with your fore finger, and place your little finger at one third part of the diftance between your fore finger and the bridge, you have G fharp.

A fharp, C, C fharp, D fharp may be made on the third ftring in the fame manner as D fharp, F, F fharp, G fharp were made upon the fourth; and in a fimilar manner you may proceed to fill up the vacancies in the remaining part of the fcale, the particular directions for which would be needlefs, after having explained fo fully thus far. *The ⅓ ftop always making the harmonic octave fifth to the note made by the fore finger in the common way.*

IT is fomewhat difficult for thofe whofe fingers are fhort, to command the ⅓ ftop to advantage. In that cafe, the ¼ ftop may be ufed as follows; but the fcale can only be completed from tne fecond G upwards.

MAKE common G sharp on the silver string
with your fore finger, and at the same time lay
your little finger as lightly as possible on the
the string at ¼ of the distance between your
fore finger and the bridge, you have the harmonic G sharp.

MAKE the common A sharp, C, and C sharp
with your fore finger, and place your little
finger at one fourth of the distance between
your fore finger and the bridge as lightly as
you can, you have the harmonic A sharp, C,
and C sharp. And in the same manner you
may proceed with the other strings. For ob-
serve, that *whatever note you make with your*
fore finger in the common way, by laying a
finger lightly on the string at one fourth of the
distance between your fore finger and the bridge,
you make the harmonic double octave to that note:
which rule is perfectly plain and easy for practice. The sounds, however, are not quite so
fine as those made by stopping at ⅓; and in neither case are they so fine as when made by the
open string, without the use of the fore finger,
except in the instances mentioned before; for
which

which reafon the notes in the fcale above fhould
be ufed whenever they can.

N.B. If the ftring be preffed down, not with
the flefhy part of the fore finger, but with the
nail, the founds will be much better. Alfo *in
general* if, when you have ftruck a note, the
finger which makes the light ftop be taken off
from the ftring together with the bow, the
found will continue a while after, and therefore be more pleafing; in the open ftring efpecially this has a fine effect when properly exe-
cuted. But if this rule be obferved only when
you ufe the hard ftop, and not when you ufe
the open ftring, the founds will be brought to
an equality of fweetnefs ; at leaft a good per-
former will be able to do it fo well, that the
difference fhall not be fenfible to an ordinary
ear. The rule, however, may be obferved to
great advantage in the following fcheme, where
only open ftrings are ufed. *(That fcheme, and
the perfecting of the former one by means of the
hard or fore finger ftop, are the two particulars
which I could not find in Rouffeau, as mentioned
before.)*

A De-

A DESCRIPTION of an HARMONIC VIOL; the Scheme of which may be applied to any inftrument of the viol kind.

THE SCHEME.

Fourth String	1	5	8	$10^{Sh.}$	12	$14^{Fl.}$	15	&c.
Third String	2	6	9	$11^{Sh.}$	13	$15^{Fl.}$	16	&c.
Second String	3	7	10	$12^{Sh.}$	14	$16^{Fl.}$	17	&c.
Firſt String	4	8	11	$13^{Sh.}$	15	$17^{Fl.}$	18	&c.
Diviſions from the nut	$\frac{1}{2}$	$\frac{1}{3}$	$\frac{1}{4}$	$\frac{1}{5}$	$\frac{1}{6}$	$\frac{1}{7}$	$\frac{1}{8}$	&c.

WHEN you are playing the violin, or other viol in the common way, and would introduce at times the harmonic notes, you muſt do it according to the directions already given. But for playing a piece all through in harmonics, you may uſe the above ſcheme, *the ſtrings of the inſtrument being tuned each one note above another*. The notes will then lie in a very natural and eaſy order for playing; and the ſtrings being open, you may manage theſe ſounds to the greateſt advantage. You may tune it to any inſtrument or pitch at pleaſure; and you may alſo flatten or ſharpen any of the ſtrings anſwerable to the key, only remembering that

all

all the notes on thefe ftrings are then flat or
fharp; and as moft of the notes are double,
you cannot be at a lofs for the natural ones, &c.
Imagination alfo may make this fcheme ftill
more complete : if, for example, you pitch in
8, you may fharpen the fecond ftring, if a fharp
key, and fuppofe the notes on $\frac{1}{5}$ and $\frac{1}{7}$ out of
the queftion, fo that the notes you want will
run on in a more eafy and natural order, and
the fharp ftring will alfo give the fharp thirds
and fevenths all through. The like of other keys
or pitches. In fome pitches you may take only
three ftrings, and tune the other a fourth, fifth,
eighth, a flat, or fharp, or whatever you have
occafion for. Thus, if you pitch in 6, you
may tune the fourth ftring a fourth under, by
which means you not only have that fourth
more convenient than by going down to $\frac{1}{2}$, but
have alfo the octave below the key, with other
notes above, which fome performances might
require : or you may add a fifth, or fixth ftring,
and referve them as by-ftrings, for thefe and
the like purpofes.

ONE finger can very eafily manage the notes
on each fret or crofs line, as the ftrings are not

to be preffed down, but the finger flipt as
lightly over them as poffible. The ftrings
fhould be all of a fize, or nearly fo; not fmall;
and as even and clear toned as poffible. If the
inftrument was longer than a violin (I mean
on account of the ftrings), and if it was made
fomewhat like a mandoline or guittar, perhaps
the founds would be more melodious ; fuch an
inftrument would do very well to play harmonics all through with ; and a mafter would not
be at a lofs to play with it by turns (by means
of fhifting) in the common way alfo.

———————

P. S. SINCE the note in page 69 was fent to
the Printer, I have fatisfied myfelf that the
harmonic founds which arife by bowing, depend entirely on the bow, as therein obferved;
for.

I. No fuch founds ever arife by making the
ftring found by any other method that I can
difcover.

II. THE founds which arife depend entirely
on the part of the ftring bowed upon : and the

F   part

part of the ſtring to be bowed on, in order to
produce any given harmonic ſound, may even
be *calculated*, by proceeding on the data delivered in the ſecond and third cautions in the
theory above.

III. The bow therefore acts in a double capacity in theſe caſes, both ſtopping, and vibrating the ſtring, as before obſerved.

IV. The eye can very eaſily diſtinguiſh when
a ſtring vibrates harmonically, and when only
in the common way; in the latter caſe, the
whole ſtring freely and viſibly forms the harmonic curve; in the former only its aliquot
parts. Both theſe caſes may indeed happen together, as hath already been noticed, but then
the latitude of the vibration of the whole ſtring
is proportionally and even viſibly affected.

V. If Tartini's theory were juſt, the ſtronger
the ſtring was made to vibrate, the louder
would the harmonic ſounds be excited: but the
contrary of this obtains; for in order to raiſe
theſe ſounds we muſt bow very lightly, for if
the bowing be ſtrong no ſuch ſounds are heard.

I COULD

I COULD enter into a more ample refutation of this theory; but thefe few hints will be fufficient to the philofopher and mathematician. If any one choofes to examine the matter by experiment, he will do well to obferve that bowing with a common bow, and with a fingle hair makes a very material difference; the former occupies a greater fpace on the ftring, and therefore raifes more notes, and in a more irregular manner: but this is avoided by ufing only a fingle hair.

I WOULD wifh, however, not to be mifunderftood. That a ftring of an harpfichord, &c. when founded affects all thofe ftrings that are concords, I by no means deny; they do it on the fame principle that one ftring excites another which is in unifon with it, and which is too well known to philofophers to need explanation. But that a ftring when founded raifes alfo the various harmonic founds which that ftring yields by vibrations in its aliquot parts, is, I think, fufficiently refuted by what has been faid, or at leaft could be refuted by purfuing thefe hints.

AN

AN

INQUIRY

CONCERNING

COMBUSTION:

SECTION I.

The principal Phenomena of incombustible Bodies.

I. IF an incombustible body be expofed to the focus of the fun's light collected by a burning glafs, to a culinary fire, to friction, or the like, it will become hot; and its heat will be greater, according to the power of the agent. The heat will continue as long as the caufe continues to act.

II. BUT if that caufe be removed, the body lofes its heat by degrees, till it becomes of an equal temperature with the fubftances around.

F 3

III.

III. Bodies are expanded by heat, and contracted by cold; and different bodies in a greater or lefs degree, according to their denfity, the cohefion of their particles, and other circumftances.

IV. If a folid incombuftible body be heated, and another be applied to it cold, the former will communicate heat to the latter, and if the heat of the former be fufficiently kept up, it will in the end caufe the latter to be hot to any poffible degree.

N. B. By *cold*, I mean a degree of heat lefs than that of the common temperature; and by *heat*, the contrary. But it is more philofophical to ufe only the term *heat*, and to confider bodies as more or lefs hot according as they raife or fink the fluid in the thermometer.

V. When an incombuftible body is heated to a proper degree, it emits light, fo as to caufe the body to appear luminous to the eye, the light increafes with the heat; but if it be fuffered to cool, the light decreafes again with the heat; and when it arrives at

about

about the fame degree as when it began to
fhine, the light ceafes to be vifible: if another
body be applied to this when fufficiently luminous, it will alfo acquire from it a luminous
heat.

VI. BODIES heated till they become luminous, are faid to be *ignited.*

VII. SOLID bodies are rendered fluid by
heat; and fluid bodies with fufficient degrees
of heat are turned into vapour. But different
degrees of heat are requifite to produce thefe
effects on different bodies.

SECT.

SECTION II.

The Phenomena of combuſtible Bodies.

HAVING premiſed as much as was judged neceſſary concerning the heat and light of uninflammable bodies, we may proceed to the ſubject of inflammable ones.

I. If a ſufficient heat be applied to a perfectly inflammable body *expoſed to the air*, it will kindle into a flame: but this flame does not require the aſſiſtance of the cauſe by which it was kindled in order to the continuance of its heat and light, as is the caſe with inflammable bodies. It has the property of maintaining or ſupporting itſelf till the whole of the body or ſubſtance is conſumed.

II. A body, or vapour, when heated as above is ſaid to be *red hot;* but the adjective ſhould be varied: and we may with equal propriety ſay that the flame of ſulphur is *blue hot*; the flame of copper *green hot*, and ſo of other co

lours

lours. Uninflammable bodies, in the firſt degree of luminous heat, emit the red-making
rays moſt copiouſly, and thence are ſaid to be
red hot. If they are heated more violently,
they emit all the rays in more equal proportion, and thence are ſaid to be white hot. Thus
alſo the flame of ſulphur emits the blue-making
rays moſt copiouſly ; the flame of copper the
green, and ſo of others ; *ſhining hot*, therefore,
would be a more proper general expreſſion.

III. In order that combuſtible bodies may
burn, or flame, they muſt not only be expoſed
to the air, but raiſed into *vapour* : and even the
vapour thus raiſed muſt be put into a proper
ſtate, otherwiſe no flame will be produced.
Thus, ſpirit of wine may be evaporated entirely
in open air, and yet no combuſtion happen.

IV. The chief circumſtance requiſite to the
inflammation of a combuſtible vapour in open
air, is *a due degree of heat* ; if that be applied
to the vapour when properly compreſſed by the
atmoſphere, it inflames, in whatever manner
the heat be communicated. The touch of an
inflammable body already burning, or of an uninflammable

inflammable body ignited, is not neceſſary for that purpoſe.

V. DIFFERENT combuſtible bodies require different degrees of heat to make them burn ; for as they only burn by a flame, they muſt firſt be raiſed into vapour. But different inflammable ſubſtances require different degrees of heat to raiſe them into vapour according to their volatility : and even afterwards, this vapour is more or leſs difficult to be turned into flame, according as it is in its nature more or leſs combuſtible.

VI. IN order that the combuſtion may be continued after once begun, without the aſſiſtance of extraneous heat, the body muſt be poſſeſſed of a ſufficient quantity of the inflammable principle, or *phlogiſton;* and then, if the other ingredients of that body be in due proportion, and ſufficiently volatile, the combuſtion will continue as long as any of the ſubſtance remains; as happens with alcohol. If the phlogiſton, though in ſufficient quantity, be combined with matter of a fixed nature, the aſſiſtance of extraneous heat is neceſſary to the combuſtion,

as

as without it the particles of the body with which the phlogifton is combined, cannot be duly expofed to the action of the air : this happens with fome metals. Vegetables, and moft other combuftible bodies partake of both thefe cafes. And even after the latter operation is carried as far as poffible, a fubftance will remain which is a truly incombuftible body.——The combuftion in the former cafe is called *inflammation ;* in the latter, *calcination.*

VII. A s ʜɪɴɪɴɢ heat in the body of the matter to be burnt has, properly fpeaking, no connection with its combuftion. Thus iron is ignited before its combuftion begins ; fulphur, on the contrary, burns before it has acquired that degree of heat. Burning phofphorus cannot fet fire to zinc ; but zinc can inflame phofphorus long before it has acquired even a luminous heat. Different fubftances require different degrees of heat to begin their combuftion, as mentioned before : and if the due degree of that heat be applied, provided the vapour be fufficiently inflammable, duly condenfed, and expofed to the action of the air, the inflammation takes place, though the body

by

by which it is communicated be neither in actual combuſtion, nor ignited.

VIII. It has been ſufficiently demonſtrated by philoſophers that combuſtible bodies contain a principle which *they call phlogiſton;* and that this conſtitutes the eſſential difference between combuſtible and incombuſtible bodies ; I ſay which *they call* phlogiſton, for they ſuppoſe that this principle is reſolved into elementary fire by combuſtion, and hence they account for the *heat* and *light* attending this procefs : Dr. Black terms it, for this reaſon, *the principle of inflammability,* and others again, *the inflammable principle.* But it will appear, in the courſe of the following Eſſay, that the phlogiſton is a fixed principle, of a nature very different from what it has hitherto been imagined ; that it is not fire ; and that it is only *mediately* the cauſe of heat in combuſtion. For theſe, and other reaſons which will be ſeen in the ſequel, I would ſubmit to the learned whether any of the terms above mentioned ought to be continued ? and whether *electron,* or ſome other, ought not to be ſubſtituted in their ſtead ? I have, however, uſed the old word, till I have their approbation for adopting a new one.

SECT.

SECTION III.

Of the Principle on which Combuſtion depends.

EXPERIMENT I.

IF alcohol be evaporated with an heat not ſufficient to inflame it, and the vapour be condenſed, it will be found the ſame ſubſtance as before.

Exp. II. If the vapour of alcohol be inflamed, and what flies off condenſed, it will not be found to be alcohol, nor even an inflammable ſubſtance; for nothing but water can be diſcovered in it.

Corollary I. By inflammation, therefore, the vapour of alcohol is decompoſed: and this holds good with all inflammable vapours.

Exp. III. If the wick of a candle be ſet fire to in open air, the flame will continue until the candle is burnt.

Exp.

Exp. IV. But if it burns only in a certain quantity of air, the combustion will continue only during a time; which will be greater according to the quantity of air. If this air be exchanged for fresh, and the candle again lighted, it will burn only about the same time as before. By changing the air a sufficient number of times, the whole candle may be burnt out as completely as if it had not been confined in a close vessel: but no art can continue the combustion without such renewal of the air.

Corol. II. The second experiment shewed that inflammable vapours are decomposed by combustion, and reduced to the state of *uninflammable bodies*. They were, therefore, decomposed by having their *phlogiston* taken from them. In this experiment, we find that the air takes something from the burning vapour; for after a vapour has burnt in a given quantity of air during a sufficient time, the combustion cannot any longer be continued; though if fresh air be added, it may; the air therefore was saturated with something which it had taken from the inflamed vapour; but what the

vapour

vapour loft was the principle which conftituted it an inflammable fubftance. It was the *phlogifton* therefore which the air took from the vapour, and with which, in the end, it was faturated. Now, as the flame continued only while the air was taking the phlogifton from the vapour, and went out when the air was no longer able to do this, it feems " that the combuftion depended entirely on fuch action of the air on the phlogifton."

Exp. V. If a bit of charcoal be inclofed in a large veffel, and made fufficiently hot, and then the whole be fuffered to cool, the air in the veffel will be found fatured with phlogifton. If frefh air be added to the coal (the firft being withdrawn), the operation repeated, and fo on fucceffively for a number of times, the phlogifton of the coal will be very fenfibly diminifhed, as I have tried. And, therefore, if the operation had been repeated a fufficient number of times, the whole of the phlogifton might have been extracted as completely as if it had been burnt in the open air.

Corol. III. A coal is a combination of phlogifton with earth; but by this experiment

it

it appears " that the phlogifton has a greater affinity with air than with the earth of the coal; and therefore when the proper circumftances concur, it quits the latter to join with the former." The circumftances which attend this are fimilar to what happens in other chymical decompofitions. If I put a quantity of fixed alcali united with fome other fubftance, fuppofe fulphur into a glafs, and pour on it a little vinegar, the vinegar will extract the alcali from the fulphur until it is perfectly faturated therewith; but even if heat be afterwards applied, it will not extract any more; neither will air, when faturated with phlogifton, extract any more of that principle from the charcoal. If now the faturated acid be feparated, and frefh poured on, more alcali will be taken from the compound; and thus we may proceed till the whole is drawn out; the fulphur will then remain behind, deprived of its alcali, in the fame manner as the earth of the coal remained behind deprived of its phlogifton. The ftrength of the analogy will eafily be perceived by the philofophical reader.

THE air faturated as above is called *phlogifticated* and *fixable air*.

Exp.

Exp. VI. If in open air any heat be applied to *sulphur* below a certain degree, it will not not burn. If fulphur be inclofed in a veffel with fixed air, and a greater degree than that with which it would be burnt in the open air be applied, it ftill remains uninflamed, and the fame fulphur ; but if the veffel with the fulphur in this ftate be uncovered, it kindles into a flame immediately on the admiffion of air, without the application of a body already burning, and is entirely decompofed.

Corol. IV. The fubftance which has hitherto been confidered as having one of the greateft degrees of affinity with phlogifton is the vitriolic acid ; for moft other fubftances which contain that principle, part with it to this acid, when the circumftances requifite to their union concur ; the fubftance formed by their union is *fulphur*, the fubject of the above experiment. But it appears that phlogifton has a greater affinity with *air*, than with *vitriolic acid :* for, when the proper circumftances concur, it quits the latter to join with the former. The affinities of phlogifton therefore, with refpect to thefe fubftances, fhould be placed thus:

G   Phlogifton.

Phlogifton,

Air,
Vitriolic Acid,
&c.

Now that the phlogifton is really united with
the air, by means of a fuperior affinity, as explained above, appears not only from what has
been faid, but alfo from this confideration, that
the air thus combined is altered in its fpecific
gravity, is lefs elaftic, and in other refpects
changed in its properties. The properties of
the vitriolic acid are likewife altered on its
combination with the fame principle, with an
alcali, or any other fubftance. The faturation
of air with phlogifton, is as analogous to the
faturation of the vitriolic acid with the fame
principle as any two proceffes of the kind can
be, allowing for the very different natures of
thefe fubftances ; but the following analogy
will fet it in a ftill clearer light. Common falt,
and cubic nitre, may, in this view, be confidered as fimilar, except in the attraction which
their alcaline bafes have with their refpective
acids. The vitriolic acid decompofes the former with greater eafe than the latter, becaufe

its

its principles are united by a weaker attraction. In like manner, phofphorus and fulphur may be confidered as differing from each other only in the affinity which the phlogifton has for the refpective acids. But the air decompofes phofphorus with greater eafe than fulphur: and for this no other reafon appears but that the phlogifton has a weaker affinity with the phofphoric than with the vitriolic acid. *In the procefs of combuftion, therefore, we muft reafon in the fame manner as on other chymical affinities and decompofitions.*

SECT.

SECTION IV.

Of the Phlogiston.

THE doctrine of combustion which at pre-sent prevails is, that the phlogiston is *combined elementary fire:* that in this procefs it is fet at liberty, and refumes its elaftic ftate ; and that the heat and light of flame, as alfo the property which it has of fupporting or main-taining itfelf, proceeds from the avolation of this difengaged principle.

THE phenomena of fixed air are made ufe of to illuftrate this theory, and from hence, in-deed it was, by analogy, derived. Fixable air may be combined with various fubftances, and form with them grofs bodies : thus it may be combined with the cauftic volatile alcali ; it may be transferred from thence to other fub-ftances with which it has a greater affinity, to the cauftic fixed alcali for example ; and from thence again to quicklime. But if a fubftance be applied to the compound with which the quicklime or alcali has a greater affinity than

with

with the air, it lets go the air, and unites with that fubftance. The air thus difengaged, and no other fubftance at hand with which it can unite, refumes its elaftic ftate, and becomes permanent air; as it flies off it caufes an effervef-cence in the liquid in which it was contained.

It is fuppofed that elementary fire may, in like manner, be combined with bodies, and that it may be transferred from thefe to others for which it has a ftronger attraction: thus, it may be combined with the earth of charcoal. From thence it may be transferred to metallic * calxes; from thefe to the phofphoric acid, and from thefe again to the vitriolic. In the pro-cefs of combuftion, it is confidered as " let go by the body with which it was combined; that it refumes its elaftic or expanfive ftate, and, by its flight, produces the phenomena of heat, &c. after the fame manner as air produces ef-fervefcence." This theory is ingenious, but I think not true, for the following reafon:

It is known to chymifts, that pure fulphur is a combination of the vitriolic acid with phlo-

* Zinc, &c.

G 3   gifton,

gifton, and that thefe are the only ingredients which enter into the compofition of that fub-ftance. In combuftion the fulphur is decompo-fed by means of a third body, or air; which having a greater attraction for one of the in-gredients than that which is already combined with it, that attracted ingredient quits the other, and unites with the air. Now if the analogy held good, we muft fay " that the vi-triolic acid had a greater affinity with the air than with the phlogifton, and therefore let go the latter to join with the former; that the phlogifton thus difengaged, refumed its elaftic ftate as elementary fire, and caufed by its flight the phenomena of heat, &c. juft as difengaged air caufeth effervefcence." But the reverfe of this happens, for the acid is left behind; and of courfe " it was the other ingredient, or phlo-gifton, which combined with the air."

STRANGE as it may feem, this laft fact is acknowledged by thofe very perfons who em-brace the theory of combuftion above explained; fo that it is matter of furprife that they have not difcerned the perfect analogy of this pro-cefs with other chymical ones of the like kind.

Perhaps

Perhaps the light which attends combuſtion has *dazzled their eyes*, and proved an *ignis fatuus* which has led them out of their way. Perhaps alſo, the property which inflammable bodies have of maintaining the combuſtion once begun in them, has proved their ſtumbling block. For we ſhall find that theſe phenomena admit of an eaſy ſolution from the doctrine above laid down.

THE *deſtruction* * of the phlogiſton in this proceſs has been a favourite doctrine ſince Stahl honoured it with his opinion. But when vinegar extracts the alcali from liver of ſulphur, as in an experiment before related, chymiſts do not ſay that the alcali is deſtroyed, as they ſay that the inflammable principle is deſtroyed by combuſtion; their ideas on that head are clear enough ; they rightly conclude that the alcali had left the ſulphur to unite with the vinegar. They argue in the ſame manner on the decompoſition of charcoal by the vitriolic acid ! why this reaſoning ſhould have been departed from in the inſtance before us, is not eaſy to imagine.

* Vide Macquer's Chymical Dictionary.

IT

IT would be eafy to bring other objections to the prevailing theory of combuftion; but as I imagine that firft ftated to be an *argumentum crucis* (if the expreffion may be allowed), it will be needlefs to trouble the reader with more.

PART of this theory however is true, as will hereafter appear. The *light* of flame proceeds from the difengaged phlogifton, though the *heat* does not. Alfo the *heat* really proceeds from difengaged fire, as chymifts at prefent imagine; they are only miftaken as to the origin of that fire.

SECT.

SECTION V.

Of the Heat and Light attending Combustion.

IT is well known to chymists that when certain bodies unite, their combination is followed by a greater heat than what those bodies
possessed before. Thus heat is generated when
an acid is saturating an alcali. The like happens when water is mixed with spirit of wine;
and a still greater heat follows on mixing water
with the mineral acids. In some cases therefore the heat may perhaps be so great that the
new compound shall be luminous. When water
is mixed with quicklime, the heat is sometimes
so great as to kindle combustible bodies with
which it happens to be in contact. In the pyrophyrus of Homberg, the union of water with
the vitriolic acid is attended by so great an
heat, that the inflammable substances in the
compound are set on fire by it ; and the nitrous
acid and oils actually inflame.

THE

THE caufe of this heat I defer the confideration of to a future fection : the fact alone is fufficient for our prefent purpofe. The heat generated by the combination of phlogifton and air may, for the prefent, be reckoned analogous to thefe ; and, in ordinary combuftion, is fo great as to be luminous, as may be gathered from the fecond Corollary in the third fection hereof.

IT may farther be obferved that combuftible bodies are heated to a degree before they begin to flame : and it may be prefumed that the fame quantity of heat is generated by the combination whether the particles before their union were hotter or colder. When, therefore, the ingredients are previoufly heated, or their temperature is higher ; the heat after their combination will be greater than if they had united in a colder ftate, becaufe the heat generated by their union, is added to that which they had before acquired. If this be joined with the confiderations in the preceding paragraphs, it will ceafe to be a wonder that this procefs is attended with a *fhining heat*.

Now

Now, in cafes where the heat is not intenfe, as in the combuftion of fulphur, I apprehend that the combined phlogifton and air only are luminous : but if it be fufficient, a fhining heat will be generated in the extraneous particles of the vapour. I defer a particular confideration of the light of flame to a future fection, and fhall here only obferve that a vapour whofe particles are rendered luminous, muft appear to us under the form of *flame*.

S E C T.

SECTION VI.

Of the Continuance of Combuſtion.

IN the foregoing pages we have endeavoured to explain the principle on which combuſtion depends, and the phenomena of heat and light which attend the combuſtion of inflammable bodies. The property which theſe bodies have of maintaining the combuſtion after it is begun, ſhall be the ſubject of this ſection. I ſhall treat this ſubject in as conciſe a manner as I did the others; and an attention to the following circumſtances will ſufficiently explain my ideas on that head.

LET a vapour be raiſed from a perfectly inflammable ſubſtance in open air, let that vapour be properly compreſſed by the atmoſphere, and a ſufficient heat applied, the particles of air will attract the phlogiſton from the particles with which it was before united, and a ſhining heat will follow. The vapour therefore will appear under the form of *flame*. The par-

ticles

ticles thus ignited will be enabled to commu-
cate heat to thofe on the furface fufficient to
raife them into a vapour proper to be acted on
by the air. This vapour being in like manner
decompofed, thefe frefh ignited particles of phlogifton and air will communicate heat to thofe
next on the furface, which therefore will likewife be elevated and decompofed; and fo on
in a continual fucceffion, as long as any of the
fubftance remains: for, as by the combination
of each particle of air with phlogifton, heat
is generated, and that in the great degree mentioned above, fo many particles as thus combine,
fo many new fources or fprings of heat will
there be; which, with what follows, will be
amply fufficient to account for the phenomenon
in queftion.

For this procefs is affifted or facilitated by
the action of the atmofphere, by which the vapour is compreffed, and the particles of air and
phlogifton forced into contact: hence the combuftion goes on in an heavy atmofphere better than
in a light one: Hence alfo, when bodies burn
in clofe veffels, the flame ceafes before the air
is all faturated with phlogifton, becaufe its elafticity

ticity being weakened, the vapour is not fufficiently compreffed. Bellows, and currents of
air, befides that they drive away the faturated
air, and apply frefh particles to the vapour, af-
fift combuftion on this princlple.

IT appears, from what has been already faid,
that different combuftible bodies require different degrees of heat to make them flame. The
degree of heat therefore which is neceffary to
begin the combuftion will, for the fame reafon,
be required for its continuance; now, whatever this requifite degree of heat be, yet if the
body be perfectly and uniformly inflammable,
or burns wholly away when once kindled, it
will be found that *more phlogiſton and air are
combined in a given time;* and therefore a
greater quantity of heat continually generated,
fufficient to equal that firft degree of heat, and
of courfe to maintain the combuftion. We
have, therefore, from the above principles, a
very eafy folution of the phenomena of combuftion, and the theory will perhaps be the
more readily embraced, as its principles are
within the bounds-of common obfervation.

THE

THE reader will eafily apply the doctrine delivered to the particular phenomena of combuftion; with an inftance of which I fhall conclude this fection.

EXPERIMENT. If, inftead of air, *nitre* be mixed with a combuftible body, and put into a clofe veffel in vacuo, or otherwife, and then rhade fufficiently hot, the combuftion of that body will be as complete, as if it had been burnt by means of air.

THE nitrous acid, therefore, contains a quantity of air of the fame nature with that of the atmofphere, but in a combined ftate : as foon as the proper circumftances concur, the phlogifton in the inflammable body, and the air in the nitrous acid, by a mutual attraction are withdrawn from the fubftances with which they were before combined, and unite with a fhining heat, in the fame manner as atmofpherical air and phlogifton in the inftances defcribed. The air thus at liberty, refumes its elaftic ftate, and becomes the fame fixed air as that generated by common combuftion. The phenomena of gun-powder and other nitrous combuftions combuſtions may be underſtood by means of this theory.

CHYMISTS have obſerved, that if phlogiſton be combined with the phoſphoric or vitriolic acids, *ſulphurs* are formed : as nitrous acid has a greater affinity with phlogiſton than either of theſe, they conclude that a ſulphur is likewiſe formed by their combination. " But (ſay they) the nitrous acid and phlogiſton unite with ſuch violence, that the ſulphur is deſtroyed the very inſtant that it is formed * " Is not the fixable air produced in this proceſs the nitrous ſulphur ? and would it not be conſiſtent with chymical analogy, and therefore more proper, to call fixable air in general Aërial Sulphur ? but this is ſpoken with ſubmiſſion to better judges.

IT has generally been ſuppoſed that the great attraction which the nitrous acid has for phlogiſton, is owing to its containing that principle as *a conſtituent part* †. But the reaſon now appears to be that it contains common *atmoſphe-*

* Macquer's Chymical Dictionary.   † Ib.

rital air; I take it that it contains phlogiston in no other manner than as the volatile vitriolic acid does; and that by expofing to the air the phlogiston is diffipated; for it ceafes to fume, and becomes fixed like the oil of vitriol. Perhaps air conftitutes the *effential part* of the nitrous acid, on which its tafte, corrofivenefs, and other general properties depend; and it feems to me that it is combined with nothing but water by means of an earth : for water only is to be found in the nitrous clyffus, and and the earth may be left behind with the alcali. I ufed to think that it was combined with water alone: but if the reafoning in a following fection on air be admitted, that cannot be the cafe; for water parts with air with a lefs degree of heat than nitre does, and therefore there muft be fome other fubftance of a more fixed nature with which it is combined, and only mediately with the water by means of that fubftance. Is the effential part of the *vitriolic acid* alfo atmofpherical air in a ftate of combination *, but combined with fuch fub-

* Do not the explofions which have been obferved of balfam of fulphur favour this opinion? did the air and phlogifton mutually difengage each other, and form fixed air?

H   ftances

ftances, or in fuch a manner, that it is not fe-
parable by the phlogifton, as in the nitrous?
and is this the reafon of its great affinity with
phlogifton? and may the like query be put
concerning the other acids? but this by way
of digreffion.

FROM what has been faid, it appears that the
phenomena of combuftion depends on this principle; that air has a greater affinity with phlogifton than the fubftances have with which it
is combined in inflammable bodies, and therefore when all circumftances properly concur,
it attracts that principle from thofe bodies;
that a fhining heat is generated by their combination; and that this decompofition when
once fufficiently begun in a perfectly inflammable body, together with the fhining heat
which is a confequence thereof, will be continued on the principles above laid down, without any farther affiftance from extraneous heat,
as long as any of the fubftance remains.

WHEN phlogifton is combined with the earth
of charcoal, with the vitriolic acid, or certain
other fubftances, a combuftion may happen by
reafon

reafon that the air can attract it from thefe
fubftances; but when it is combined with air
no combuftion can happen, becaufe air cannot
attract phlogifton from other air, any more
than the vitriolic acid can attract it from fulphur ; for the affinities are equal, and one fubftance cannot attract another from a third, with
which it is combined, but by means of a fuperior affinity. This may alfo be applied to the
nitrous fulphur above fpoken of.

THERE are however, certainly, fubftances
capable of attracting the phlogifton from air,
otherwife the whole atmofphere would in time
be converted into fixable air. The ingenious
Dr. Prieftley, to whofe labours the learned
world is fo greatly indebted, has already difcovered two of thefe means : he fhews that
fixed air may be rendered pure by vegetables
growing in it, and by water. It may perhaps be
added, that as fixable air has a greater fpecific
gravity than common air, and therefore naturally
tends downwards, it enters into the earth, and
combines with fuch bodies as may be difpofed
to receive it. That fome fubftances may have
the property of depriving it of its phlogifton,

which then enters into their compofition; and perhaps, in fome cafes, there may be no other method of combining that principle with bodies, at leaft in certain manners. The phlogifton, therefore, when combined with air, feems to be in the moft proper ftate for certain intimate combinations of it with vegetable and other fubftances; as it is probably reduced to its integral parts. Hence we have fome idea not only of the manner in which fixable air is deprived of its phlogifton, but alfo of the circulation of the laft mentioned principle, from bodies to air, and from air again into bodies. Perhaps alfo the fixed air, when deprived of its phlogifton, may, in fome cafes, be converted into acids, if the above queries concerning acids be true; but thefe things remain to be inquired into.

SECT.

SECTION VII.

A Speculation *.

AIR has ufually been reckoned a fluid *fui generis*, and called, in contradiftinction to others which are coherent, an *elaftic fluid*.

I HAVE long been of opinion that the elafticity of air depends on heat; for if the heat be increafed, the elafticity is increafed; if it be

---

* In the Effay on the fenfes I allotted a fection for fuch hints and conjectures as had occurred to me on thofe fubjects, in order to their being further inquired into by others: I fhall devote this fection to a like purpofe. As I do not pretend to offer thefe conjectures as demonftrated truths, any errors will be pardoned by the candid, as the detection of them by experiments may lead to real difcoveries; I have, for my amufement, carried the ideas contained in this fection, as well as others which are not mentioned, to greater lengths. If this fhort extract be approved, I may in a future edition render it more copious. Thofe, however, who do not relifh fpeculative reafoning may pafs over this fection.

diminifhed,

diminished, the elasticity is also diminished in a certain regular proportion : it should seem, therefore, that if air was entirely deprived of heat, its particles would become coherent.

WHEN water is raised into vapour it is also elastic, and its elasticity is greater as more heat is afterwards applied. The vapour of water, therefore, is of a similar nature to air : the only difference, in this respect, between them is, that water requires a vast deal more heat to render it elastic. The like may be observed of other bodies *.

IF we imagine two particles in contact, and that heat be applied, the heat will force them to quit each other. The particles, while within the spheres of their cohering forces, will resist the action of heat more, as their cohering forces are stronger. But when the heat becomes so great as to force them beyond those spheres, they will be elastic, like air; their elasticity will be greater as the heat is increased, and that in a certain regular proportion; as mentioned above.

* It seems to me that this proposition may be made as general as Mr. Braun's concerning the fluidity of bodies.

CASE I. Heat, therefore, is the caufe of the repulfive affection among particles of air; and if this caufe be removed, the particles have no fuch tendency.

CASE II. Fire furrounds a particle of air in the manner of an atmofphere; it is denfer near the particle, and rarer at a diftance from it; and hence the repulfive power of particles of air. Fire therefore is attracted by the particles of air.

CASE III. When air is hotter it is more elaftic; that is, its particles are furrounded with greater and denfer atmofpheres of fire, and therefore their repulfive powers are ftronger.

CASE IV. As our atmofphere, by its gravity, is denfer, or more compreffed near the earth, than at a diftance from it, and that in a direct proportion; fo by the gravity of fire towards a particle of air, the fire is in like manner compreffed. The denfity diminifhes according to the diftance from the particle; and hence the repulfive force of particles of air is directly as the diftance of thefe particles from each other.

CASE

CASE V. When the quantity of fire com-
pofing the atmofphere of a particle is greater,
the compreffion or denfity of the fire near the
particle will be greater, juft as happens with
our air : and therefore the repulfive force of
the particle will be increafed with the heat,
and will alfo reach to a greater diftance *.

CASE VI. If two fimilar particles of air, but
with unequal atmofpheres, be brought fufficiently near to each other, the particle which
has the greateft atmofphere will part with fire
to that which has leaft, till their quantities are
equal. The like may be obferved of other homogeneous particles,

THE reafon of this is obvious; the attraction for fire, or the gravitation of fire towards
them, being equal in both or all the particles.

CASE VII. IF by any means the gravity of
fire towards a particle, or, if you will, the at-

* It is obvious from hence, that the repulfion at the fame
diftance from the particle does not increafe in a direct proportion with the heat, but in a ratio which will eafily occur to
the mathematician.

traction

traction of a particle for fire be weakened, the atmofphere which that particle retains at the fame common temperature will be lefs in proportion thereto, and it will alfo be lefs denfe ; fo that its repulfive power will likewife be diminifhed.

The idea on which this cafe is founded was fuggefted to me by the following confideration.

It has long been known that lead by calcination acquires confiderable weight ; M. Margraaf has difcovered that the acid left behind after the combuftion of phofphorus is almoft half as heavy again as the phofphorus employed : and yet in both thefe cafes, many particles muft have efcaped befides the phlogifton. I am told that Dr. Black has made experiments on metals with acids which agree fo well with thefe that he is convinced of the truth of the inference which has been drawn from them, viz. " That the gravity of bodies is diminifhed by their combination with phlogifton."

The caufe of gravity, as conjectured by Sir Ifaac Newton, and now generally fuppofed by
philofophers,

philofophers, is a very fubtile elaftic medium, which is rarer near a particle of matter, and denfer at a diftance from it. That therefore two fuch particles will be mutually impelled by the denfer, towards the rarer parts of this medium, and in courfe towards each other.

As the æther is *rarer near*, and *denfer at a diftance from, a particle*, it fhews that there is a *mutual repulfion* between the particles of bodies and this fluid.

THAT, therefore, according as this *mutual repulfion* is greater, the rarity of the medium near the particles, and the force of the particles' gravity, will be greater; and as the *mutual repulfion* is lefs, the æther near the particles will be lefs rare, and their gravity diminifhed.

IT therefore appeared to me, that as phlogifton when combined with the particles of bodies diminifhes their gravity, it does it *by weakening the repulfion between thefe particles and æther.*

TILL now, I had imagined, with Dr. Black *,

* To fome notes which a friend had taken while attending the lectures of that great phyfician and philofopher (I wifh
they

and others, that fire, æther, and phlogiston were one and the same fluid : but on applying this reasoning to my notion concerning the re- pulsive force of the particles of air, I found that it entirely clashed with it ; for phlogiston weakens the elasticity of air. And now, for the first time, it occurred to me that the dis- position of particles of bodies towards æther and fire are quite opposite. For whereas fire gravitates towards, or is attracted by those par- ticles, æther on the contrary is repelled ; and this also led me to consider that æther causes the gravity or attraction of particles towards each other ; fire, on the contrary, their repul- sion. It followed, therefore, that if phlogiston diminished the repulsion between the particles of bodies and æther, *and thereby leſſened the mutual gravity of theſe particles;* it muſt on the contrary *diminiſh the attraction between theſe particles and fire, and of courſe weaken their mu- tual repulſion.*

they had been more perfect), and to extracts from Dr. Prieſt- ley's diſcoveries concerning air, I owe my having been en- abled to work this latter part of my Eſſay into its preſent form.

THIS

THIS fuggefted to me another idea. If particles, when combined with phlogifton, have their attraction for fire diminifhed, it fhould follow, that the fame quantity of fire added to a phlogifticated, and an unphlogifticated particle, would heat the former moft; becaufe it would be lefs forcibly retained by that particle than by the other, and therefore a greater quantity would be communicated to a third and colder particle applied. To fatisfy myfelf with regard to the juftnefs of this idea, I made the following experiments.

EXPERIMENT I. In an iron pot filled with fand, heated over the fire, I placed, at a fmall diftance from each other, two gallipots, one of which contained water, the other lamp-oil, fo that they were in equal degrees of heat; after they had remained fome time, I found that the oil had acquired a greater heat than the water.

Exp. II. I made the fame experiment with minium and lead, and found that the metal acquired a greater heat than the calx.

EXP.

Exp. III. I tried the fame with feveral other fubftances, whofe fpecific gravities would admit of the experiment, with the like refult.

I ENDEAVOURED to make the experiment with common and fixable air; but the fluids being fo rare, and not properly confined, and the heat of the containing veffels fo much interfering, I could do nothing to my fatisfaction; and therefore could only infer from analogy; for the only thermometers that I ufed were conic fcurvy-grafs phials, and Daffy's elixir bottles, with fpirit of wine in them, the furfaces of which I marked on the outfide with ink. The three firft experiments however, and the confideration that fixed air is lefs elaftic than common air, feemed fufficiently to eftablifh the truth of the propofition which they were intended to determine.

So then inftead of æther, fire, and phlogifton being the fame, as is at prefent fuppofed, they appear to be three diftinct and very different fluids; and their relations or affections towards the particles of bodies and each other, feem to be as follows.

I. SINCE

I. Since when the particles of bodies contain leaft phlogifton, fire gravitates towards them moft, there is a mutual attraction between the particles of bodies and fire.

II. And fince when the particles of bodies are moft free from phlogifton, æther avoids them moft, there is a mutual repulfion between thofe particles and æther.

III. Phlogiston and æther therefore mutually attract each other; and on the contrary.

IV. There is a mutual repulfion between phlogifton and fire.

Phlogiston therefore is to æther, what the particles of bodies are to fire: for as fire proceeds from denfe to rare in its progrefs from the latter, fo æther proceeds from denfe to rare in its progrefs from the particles of the former. It feems then that there are two different, and as it were oppofite kinds of fubftance of which bodies are compofed, and two elaftic fluids anfwerable to them. And if this be true, we muft reckon four general principles, viz. *æther, fire, phlogifton,* and *the particles of bodies;* but as this laft name is too general, fuppofe we call what is meant by them *earth*. Perhaps, however, æther and phlogifton may be compofed of the fame matter, and fo may fire, and earth: perhaps all four may be only different modifications of the fame fubftance; for we muft go beyond fire and æther before we arrive at the ultimate principles of nature. There may be ftill fubtiler principles on which the elafticities, and other properties of thefe depend.

THESE four fubftances may perhaps be confidered as the *four elements* with more propriety than thofe of Ariftotle which fo long prevailed, and perhaps the phenomena of nature may be better underftood by means of them than they are at prefent.

THE proportion of earth in nature feems to be much greater than that of phlogifton, and the proportion of æther much greater than that of fire: the gravity of bodies towards the earth, I think, proves this.

ALSO phlogifton appears to be much more fubtile than earth, and æther than fire; for the
elaftic

elaftic fluid formed by particles of earth with
atmofpheres of fire, is much groffer than that
formed by particles of phlogifton with atmofpheres of æther. Thus, light is more fubtile
than air, and electricity than fixable air ; the
analogy between thefe may hereafter be fhewn.

PARTICLES of phlogifton freeft from earth
attract and retain the greateft atmofpheres of
æther; and particles of earth freeft from phlogifton attract and retain the greateft atmofpheres of fire. Hence particles of earth, when
moft free from· phlogifton, are moft elaftic or
repulfive ; and fo are particles of phlogifton
when freeft from earth.

THOUGH earth attracts fire, which is repelled by phlogifton, yet there is a ftronger attraction between phlogifton and earth, than
between any other two of the principles.

ONE particle of earth cannot cohere with
another, unlefs one or both be previoufly combined with a fufficient quantity of phlogifton.
The phlogifton both attracts the particles of
earth, and difperfes their atmofpheres of fire,
which

which kept them afunder *. Earth, for a like reafon, is the principle of cohefion between the particles of phlogifton.

SUPPOSE a particle of earth, and another of phlogifton, with each its proper atmofphere of fire and æther, if they could be forced into combination, they would quit their attractions for æther and fire, and exert their forces on each other; or they would lofe fo much of their attractions for thofe mediums (and therefore of the atmofpheres) as they exerted on each.

---

* If phlogifton be fuppofed to compofe the cohering forces of the particles of earth, the ætherial atmofpheres of the particles of this phlogifton, though greatly decreafed by being combined with earth, will yet extend to a little diftance beyond the cohering atmofphere fufficiently ftrong for producing a fenfible effect, and will furnifh us perhaps with the caufe of the repulfive force obferved by Sir Ifaac Newton, viz. that two object glaffes will lie on one another without touching; that two polifhed marbles are with difficulty made to cohere; that beyond the cohering forces of bodies there is a repulfion; that the rays of light are inflected, reflected, and refracted by bodies, and the like. The lefs forcibly the phlogifton is combined, the greater muft thefe effects be.

I   BUT

BUT pure earth and phlogifton cannot directly combine, by reafon that their atmofpheres hinder their union. Thus light cannot be combined directly with air; but if the light be prefented to a particle of earth already combined with a proper quantity of phlogifton, whereby its atmofphere of fire may be fufficiently diminifhed, the light can enter into combination with it, and then air, by reafon of a fuperior attraction, can take it from that particle. The like may be obferved of phlogifton previoufly combined with earth: The folution of fome curious phenomena feem to depend on this principle, as may hereafter be fhewn, if this fpecimen be approved.

IT may be proper to obferve that there is not a perfect analogy between earth and æther, and phlogifton and fire, as may at firft view be imagined. The quantity of fire in the univerfe feems to be very fmall, and to be only confined to the planets and other heavenly bodies; round the earthy particles of which it forms atmofpheres, as hath been defcribed, and perhaps of no very great extent even in particles of air. But æther is in quantity vaftly fuperior; and

as

as its great fuppofer fays, "is expanded throughout all the heavens;" particles of earth therefore will have, befides their limited, decreafing,
or repulfive atmofpheres of fire, increafing or
attractive atmofpheres of æther, extending perhaps to the utmoft bounds of the univerfe. But
particles of phlogifton will have decreafing or
repulfive atmofpheres of æther, reaching to the
fame diftance as thefe laft, but no increafing or
attractive atmofpheres of fire, or at leaft only
momentary ones; becaufe, on account of the
fmall quantity of this medium, and its not fill-
ing the univerfe, it will all be attracted by, and
gathered about the particles of earth.   To illuftrate this it may be noted, that when to a
particle of earth another of phlogifton is added,
part of its repulfive atmofphere of fire is diflodged. Now, if the analogy held good, the
diflodged fire ought to go into the increafing or
attractive atmofphere of fire of the particle of
phlogifton added; but, on the contrary, it goes
into the repulfive atmofpheres of fire of the
particles of earth around, as is proved by the
thermometer and the fenfe *. On the contrary,
when to a particle of phlogifton another of

* Vide cafe XI.

earth

earth is added, the æther which is expelled
from the repulſive atmoſphere of the former,
goes into the attracting atmoſphere of the latter.
For whenoil of vitriol is mixed with water,
fire is diſlodged from the particles of earth; and
by the ſame reaſon æther muſt be diſlodged
from the particles of phlogiſton. Now heat is
cauſed by the diſlodged fire, becauſe it goes into the repelling atmoſpheres of fire of the ſur-
rounding particles of earth, as obſerved above *.
But if the ſame rule held good with the diſlodged æther, the mixture would weigh heavier
than the ingredients did before, becauſe the
particles of phlogiſton having leſs repulſive atmoſpheres of æther, would be leſs repelled by the
globe of the earth, and therefore they would
have leſs levitation, or centrifugal force. But
the æther diſlodged from the repulſive atmoſpheres of the particles of phlogiſton, goes
into the attracting atmoſpheres of the particles
of earth with which the phlogiſton is combined,
and which therefore by that combination had

---

* Perhaps Phlogiſton does not repel fire atmoſpherically,
but by particle and particle, their ſubtilty being nearly alike.
Or do they really *repel*, or only *expel* one another, as the fixed
alcali expels the volatile from acids?

its

its repulsion for æther diminished; and therefore
what the particles of phlogiston lost in centrifugal,
those of earth lost in centripetal force, so that
the weight continued the same *. When these
ideas first occurred to me, I made experiments
with oil of vitriol and water, and with spirit of
nitre and ice, to see whether they altered in
weight after mixture. By the inaccuracy of my
weights I had like to have fallen into an error,
for the vitriolic mixture seemed heavier, and
the nitrous lighter than their ingredients; but
by repeating the experiment I discovered the
cause to be in the weights. The absolute gravities of the compounds were the same as those
of their ingredients; and consequently the de-
crease or increase of attraction of æther by the
particles of phlogiston, was balanced by an

---

* Imagine a particle of phlogiston where gravity is -1,
and another of earth whose gravity is -2. If they are combined, their gravity will be equal to the sum of their gravities before combination, or $-\frac{1}{3}$: and this will be the case
whether their combination be more or less intimate; and
whether free particles with their full atmospheres be supposed, or particles already combined; for on their separation
from their previous combinations, they will instantly acquire their proper ætherial atmospheres, as is obvious from
what has been said.

I 3   equal

equal increafe or decreafe of the repulfion of that medium by the particles of earth.

THE principles which have been propofed are, probably, of general extent. All bodies may be compounds of phlogifton and earth, with regulating atmofpheres of æther and fire ; and all the differences in thefe bodies may arife from the different proportions and manners of their combination. A new field of fpeculation feems therefore to be opened to philofophers by this theory.

CASE VIII. But if the mutual attraction be increafed, the atmofphere of the particle, and alfo its repulfive power will be augmented.

CASE IX. Suppofing two particles of air A and B, and that the gravity of fire towards B is decreafed, fo as to be but half of that towards A; the atmofphere of A will contain twice the quantity of fire of the atmofphere of B. If thefe two particles be brought near to each other, the atmofpheres will not become equal, as in cafe VI. but each particle will retain its atmofphere as before.

THE

THE reafon of this likewife is obvious; the attraction of A for fire being double that of B, and their repulfive powers will be different.

CASE X. The fame things being fuppofed, the heat of B will be equal to that of A, notwithftanding it contains but half the quantity of fire.

FOR thefe are the proportions which they would retain at the common temperature, or when placed near each other, as above : and this rule determines the heat, as is evident by what follows.

CASE XI. The fame being fuppofed, an equal quantity of fire added to A and B, will heat B twice as much as A; and the quantity of fire neceffary to raife them to equal heats, will be in proportion to the quantities of fire which, at the common temperature, they naturally retain. The like may be obferved with regard to cold.

FOR if to A a third particle were applied, which had but half the heat, but which would

I 4   naturally

naturally retain as much fire as A, it would take one fourth of A's fire from it, by cafe VI.

But if the fame particle were applied to B, it would take away half of B's fire to raife it to the above heat, though one third only would render their quantities as 2 to 1; as is obvious by what was faid in the IX. and X, cafes.

The mixture of bodies, which at the common temperature retain different proportions of fire, when thefe bodies are heated at different degrees, and the phenomena refulting from them, as alfo the equal affection of the fenfe, and of the thermometer by thofe different bodies when at like temperatures, may be underftood from this cafe, and thofe which precede it.

Definition I. When, at the common temperature, a particle is made to retain a greater quantity of fire than it would naturally do in that temperature, the particle fhall be hot.

Def.

Def. II. And if it be made to retain lefs
fire than it naturally would do in that temperature, it fhall be cold.

Bodies are expanded by heat merely becaufe their particles are furrounded with greater
atmofpheres of fire, and therefore repel each
other, fo that they are kept at a greater diftance than before. Cold is caufed in the fenfe
merely by diminifhing, and   at merely by increafing the quantity of fire in the part, and
therefore caufing a like contraction or expanfion of that part *. I do not therefore fee any
reafon for fuppofing either that the particles of
bodies are in a ftate of vibration when hot, or
that the particles of fire themfelves are in that
continual rapid motion which others imagine †.
If the latter was the cafe, the particles of air
ought to be exceedingly hot, by reafon of the
great and therefore condenfed atmofpheres of
fire which they contain ; but at the common

* It muft be obferved that pain, which accompanies thefe
fenfations when violent, is not to be confounded with the fenfations themfelves.

† Vide Macquer's Chymical Dictionary.

temperature

temperature they are no hotter than others which retain atmofpheres much lefs.

CASE XII. Imagine a particle attracting fire, and another which would diminifh that attraction; if they are at a fufficient diftance from each other, the latter will not affect the attraction of the former for fire, but it will diminifh that attraction more on being brought nearer; and when they meet, the diminution will be greateft of all.

CASE XIII. The fame things fuppofed, the attraction of the former particle for fire will be lefs, according as it is already combined with more of the latter.

CASE XIV. The attraction of the former particle for thofe of the latter kind will be lefs, according as it is already more faturated with them; for they will exift in the atmofphere at a greater diftance from the particle; and therefore they will alfo have lefs power of diminifhing the particles attraction for fire *.

* Vide Cafe XII.

THESE

THESE cafes were fuggefted to me by the following experiments,

EXP. I. It is well known that if oil of vitriol and water be combined, a great degree of heat is generated.

IN a fmall flender phial I put water, and added to it about an equal bulk of oil of vitriol ; the acid was poured down the fides of the phial, and remained at the bottom ; as foon as this was done, and before fhaking them together, I marked the height of the liquid on the outfide, and then well mixed the ingredients. A very great heat prefently fucceeded, and afterwards I found that the furface of the liquid was below the mark. But it could not have evaporated, becaufe it was clofe ftopped with a cork *.

EXP. II. I put fome pounded ice into a phial,

* I have fomewhere read that a drop of concentrated vitriolic acid, and another of water being put into a flender tube, penetrated each others dimenfions, fo as to be lefs in bulk. But the fpecific gravity of oil of vitriol firft led me to try the above experiment.

and

and added fpirit of nitre highly concentrated till it juft covered the ice; I immediately marked the height of the liquid; and after the ice was diffolved, found, contrary to what happened with the above mixture, that the liquid had rifen above the mark. A great degree of cold was generated by the combination.

I do not remember that there was any inaccuracy in making thefe experiments; and they feem to indicate that the heat and cold in thefe mixtures are connected with the contraction and expanfion of the compound. I at firft thought that it was from the contraction or expanfion of the body as a whole; but on confidering that ice is more expanded than the water from which it was formed, and yet that heat is generated by the congelation *, I concluded that the contraction and expanfion muft be confidered as in the particles themfelves; and that this always takes place in the particles on thefe occafions, though particular circumftances (fuch as new arrangements of the particles, and the like) may in fome cafes hinder the rule from obtaining in the whole mafs or body.

* Dr. Black.

THE

THE particles of the oil of vitriol and water therefore were, by some very powerful agent, drawn nearer to each other, so as to occupy less space than before; and as the principle of cohesion in earth, or common gravitating matter, was shewn to be phlogiston, it seems to have been effected by the agency of that principle. As the phial was corked, it did not seem likely that any fresh particles of phlogiston should have been derived from without. Besides, if that had been the case, the weight of the compound would have been lessened: but by repeating the experiment, and weighing the ingredients before mixture, and again immediately after, and suffering the whole to remain in the scale, properly suspended, till cold, I did not find this to happen. Now phlogiston must combine more firmly with the bodies according as they are already less saturated therewith * That water contains this principle in considerable

---

* Imagine a particle of earth, and that phlogiston be added to it in the manner of an atmosphere. The particles of phlogiston at the greatest distance, being less attracted, will retain greater atmospheres of æther. Phlogiston may be added to the particle till it can retain no more, by reason of the repulsion

confiderable quantity is obvious, by its affording
nourifhment to vegetables, by its being a con-
duftor of eleftricity, and alfo by an experiment
of Dr. Prieftley, in which the calx of mercury
was reduced by the phlogifton from that fluid,
and this alfo fhews that it was not contained
in a ftrongly combined ftate. The oil of vitriol, by reafon of the vitriolic acid *, is of a
more pure earthy nature, or is lefs phlogifticated ; and therefore its attraftion for phlogifton will be greater than that of water. The
phlogifton of the water will therefore be laid
hold of by the acid, and that ftill retaining the
water, a clofe and intimate conneftion, or ftrong
attraftion, or cohefion will take place between
the particles of the acid and thofe of the water,
fo that they will be drawn into leffer dimenfions. By sthi more intimate combination the

pulfion of the ætherial atmofpheres. For the fame reafon two
fuch particles, when overcharged with phlogifton, will have
their cohefion diminifhed by frefh addition inftead of increafed.
A particle thus overcharged, and another charged lefs, will
cohere more ftrongly than the two particles juft mentioned,
for the outer phlogifton of the former particle will be more
forcibly attrafted by the latter, than by any homogeneous
one, and therefore will more firmly combine with it.

* Vide feftion VI.

particles

particles of phlogifton will lofe part of their attraction for æther ; and, for the fame reafon, the earthy particles of the acid applied, will lofe part of their-attraction for fire. *The fire which thus becomes fuperabundant is, I take it, the caufe of the heat of the mixture.*

IF we imagine the oil of vitriol to be again feparated from the water, a degree of cold will be generated, equal to the heat from their mixture, becaufe the attraction of the earthy particles for fire will be reftored.

THIS, and what follows to cafe XVII. will probably explain the caufes of heat and cold arifing from chymical mixtures in general.

CASE XV. It feems therefore that " whenever heat is generated, without any addition of frefh phlogifton, it argues an increafe of attraction between the particles of phlogifton and earth, and a confequent diminution of the attraction between the particles of earth and fire." The contrary may be obferved with refpect to cold.

WHEN

WHEN Bodies return from an elaftic to a fluid ftate, or from a fluid to a folid ftate, heat is generated; and cold in the contrary cafes, as Dr. Black and others have fhewn; the above may be the reafon. And there are other phenomena of the kind which will occur to the Reader, probably depending on the fame principle.

CASE XVI. When a particle of phlogifton is combined with a particle of earth, heat is generated, if the foregoing reafoning be true; and if the combination be rendered ftill more intimate, frefh heat will arife.

CASE XVII. It follows therefore, that if phlogifton weakly combined with one particle, be transferred from thence to another, with which it may form a ftronger, or more intimate combination; the heat generated by the combination, in the latter cafe, will be greater than the cold generated by the decompofition in the former: and this difference will be greater, according to the difference of the two attractions or combinations. The contrary may be obferved of the generation of cold; and
the

the reafons are obvious from what has been
faid.

THIS cafe may obtain in fome kinds of chymical mixtures, particularly in certain folutions
of metals by acids. &c. But the former part
of it feems to me to be the principle on which
the heat in combuftion depends.

THAT phlogifton has different degrees of
attraction or forces of combination, with different particles, and with the fame particle * in
different circumftances, appears by the fol-

* Vide cafe XII. &c. It may be obferved that the attractive forces of fpherical bodies decreafe in the duplicate ratio
of the diftance. The rarity of the phlogifton therefore muft
increafe in the direct proportion of the diftance, as happens
with our atmofphere, and with the atmofpheres of fire about
particles of earth, as fhewn before. Sir Ifaac Newton has
fhewn, in the 369 page of his Optics, 3d edition, that the attraction of cohefion is as the diftance; which anfwers to the
above. Particles of earth do not feem to exert their whole
attractive force on phlogifton; but after faturation with it
they feem to have fome attraction left for fire, as water faturated with one falt can yet attract another: hence the cohering forces of bodies reach but to a given diftance beyond them.
The like may perhaps be the cafe with phlogifton.

K   lowing

lowing confiderations. After phofphorus is
burnt, if the acid be urged with a great heat, it
will give evident figns of containing phlogifton,
as Mr. Macquer obferves. Lead eafily parts
with a certain portion of its phlogifton by calcination; but retains another part very obfti-
nately. Phlogifton with pure vitriolic acid,
forms fulphur; but if water be previoufly combined with the acid, it forms an incombuftible
liquid. Light cannot form a combination directly
with air; but if it be previoufly combined with
proper matters, it is then combuftible phlogifton. Phlogifton transferred from charcoal to
zinc is much lefs eafily combuftible than while
it was in the charcoal, and many other inftances
might be produced.

PARTICLES of greater fixity or force of
cohefion (I mean homogeneous ones, and in
certain circumftances) appear to have lefs atmofpheres of fire than thofe homogeneous ones
whofe fixity, or force of cohefion, among one
another is lefs. Hence we find that the particles of water, for example, have atmofpheres
of fire fo fmall that with the ufual heat of the
air, their fpheres of repulfion do not reach beyond

yond their spheres of cohefion *. For in that
heat they will coalesce after having been rendered elaftic by a greater degree ; and the
particles of many bodies coalesce in an heat
much greater. On the contrary, air, which is
a vaft deal lefs fixed, or the cohefion of whofe
particles one with another is a vaft deal lefs, is
very elaftic with the ufual heat of the atmofphere, and even in the greateft cold we have
been able to produce. The atmofpheres of
fire therefore extend to a greater diftance beyond the cohering fpheres of the particles, and
a prodigious quantity of fire feems to be contained in the atmofphere of a particle of air, in
proportion to what is retained by one of the
fixed particles juft mentioned, infomuch that
air feems to be the great refervoir of this principle. If, therefore, we fuppofe the cohering
fpheres of particles to be compofed of phlogifton, and as phlogifton feems, by what has been
faid, to weaken the attraction of particles for
fire, we have a reafon why fixed or cohering
particles have lefs atmofpheres of fire than thofe
which are lefs fo: homogeneous particles, however, muft be underftood in thefe cafes; for

* See alfo fection IX.

with heterogeneous ones the cafe will be different *. Alfo, even homogeneous particles over-charged with phlogifton will not cohere fo well as when they have lefs, as appears from what has been faid ; and yet their attraction for fire will be lefs.

It feems to me that the fixed or unalterable particles of bodies, as of air, water, &c. † are not pure gravitating matter, or earth, but particles of earth and phlogifton combined together, in different proportions and manners, fo as that they may have, originally, greater or lefs attractions for phlogifton and fire, and repulfion for æther. What was advanced in cafe XII. &c. will hold good with any of thefe, but with different degrees of force, and the idea, properly purfued, might have its ufes. But however this be, the fact, that *moveable phlogif-*

---

* Thus air will cohere ftrongly with more fixed particles (as in nitre), though its particles will not cohere among themfelves : the reafon is plain from what was faid in cafe XIV

† I do not mean the primary or abfolutely folid particles, but thefe on which the invariable fecondary properties of bodies depend, and which experience fhews to be indeftructible. Thus water is the fame in all ages.

ton * is *more intimately combined in some bodies than in others, and that it weakens the attraction for fire in proportion to the force of that combination,* seems probable from what has been said, and will appear still more probable when we apply it to the solution of the phenomena of combustion in the next section. As I do not pretend to demonstration, and as these cases contain as much as is judged necessary to the subject which they were intended to elucidate, I shall not here pursue the idea any farther.

WHAT has been said of the attraction of fire by particles of earth in the foregoing cases, may be applied, under proper restrictions, to the repulsion of æther by the same particles, and to the attraction of it by particles of phlogiston, as is evident from the preceding discourse, and therefore I need not enlarge on it.

* I mean phlogiston which may be transferred from one body to another,

S E C T I O N   VIII.

Of the Origin of Heat in Combuſtion.

HAVING in the preceding ſeɛion given ſome conjeɛures concerning the manner in which fire is retained by the particles of bodies, and by what laws it is regulated ; we may proceed to examine how far they agree with the heat which attends combuſtion.

IT was ſhewn in the third ſeɛion that com- buſtion is a truly chymical procefs, and depends on this principle, " that the air attraɛs phlogif ton from the combuſtible body by means of a ſuperior affinity."

IN the fourth ſeɛion it appeared that when a particle of air attraɛed phlogiſton from a par- ticle of a combuſtible body, heat was genera- ted, and it was promiſed that the cauſe of that heat ſhould be conſidered in a future ſeɛion.

IT

It has appeared probable, in the courfe of the laft fection, that the particles of bodies have an attraction for fire; and that the attraction is greater, as the particles are freer from phlogifton. It further appeared, that according as the fame quantity of phlogifton is more intimately combined, it caufeth a greater diminution of the particles' attraction for fire. It alfo appeared that the particles of air have very great attractions for fire, and thereby retain vaft atmofpheres of it; but that the quantity of fire retained by fixed homogeneous particles is on the contrary very little.

Now, agreeable to the feventeenth cafe in the preceding fection, imagine a particle of the phofphoric acid faturated with phlogifton, and that it be applied to a particle of air : let the attraction of the acid for phlogifton be weak, but that of air ftrong, the particle of the air will attract the phlogifton from the acid ; and the combination will be ftronger than with the acid in proportion to the difference of the attractions; the heat generated by the combination, will be greater than the cold generated by the decompofition, for the fame reafons : in

K 4 other

other words, the fuperabundant fire of the air
will be more than fufficient to fatiate the increafed attraction of the acid, by the excefs of
the attraction of the air above that of the acid.
And this feems to me to be the manner in which
heat in combuftion is generated.

COROLLARY I. The difference of the attractions may be eftimated, by meafuring the generated heat, and attending to the following
circumftance.

COR. II. The fuperabundant fire of the air
will be attracted by, or gravitate towards, the
bodies around, in order to reftore the equilibrium. And according as thefe furrounding
bodies, or the extraneous particles, have lefs
attractions for fire, the more will they be
heated by this fuperabundant fire. Vide cafe XI.

COR. III. Suppofing a number or mafs of the
phofphoric particles, the more of them that are
decompofed in a given time, the greater will
be the quantity of fuperabundant fire generated
in that time; and therefore, fuppofing the furrounding bodies, or extraneous particles to be

the

the fame, the hotter will they be: but if they
be different, their heat will be different according to their attractions for fire, as may be collected from the eleventh cafe in the laſt
feſtion.

Now that fire is taken from the air in combuſtion appears from hence, that its elaſticity,
or repulſive power is weakened, and its dimen-
ſions decreaſed, ſo that the particles are not
kept at ſo great a diſtance from each other as
before ; for it was ſhewn in the firſt five
cafes that the particles are kept at a diſtance
from each other by means of their atmoſpheres
of fire. Mr. Cavendiſh has ſhewn, if I remember right, that the ſpecific gravity of fixable
air is to that of common air as $1\frac{1}{2}$ to 1, and yet
the abſolute gravity of the fixed air muſt have
been greatly diminiſhed by its combination
with phlogiſton, ſo that the difference of elaſticity of pure fixable and common air muſt be
ſtill greater than in that proportion*. Thoſe who
have conveniencies for making experiments
would do well to examine and aſcertain theſe
matters.

* See alſo the next feſtion.

IF we take a furvey of the different inflammable bodies we fhall find that fome of them
require a greater heat to kindle them, and
others a lefs. And therefore before a body
can be enabled to continue-its own combuftion,
fuch a quantity of its particles muft be decompofed in a given time, as will be fufficient to
furnifh a due portion of fuperabundant fire.
But, if the fubftance be perfectly and uniformly
inflammable, a fufficient number of particles
thus once decompofed, will furnifh fire enough
to equal the degree of extraneous heat firft applied, and therefore to continue the generation
of a like quantity fucceffively, by means of new
decompofitions, as long as any of the fubftance
remains. Vide fection VI.

SUPPOSING an equal quantity of fuperabundant fire generated by different bodies in the
fame time, " the heat of the flame will be
greateft in thofe whofe phlogifticated particles,
after parting with their phlogifton, have the
leaft attraction for fire;—— which contain
the feweft extraneous particles——and whofe
extraneous particles attract fire leaft."——
——The flame of fpirit of wine, when fufficiently gentle, is not even ignited, as will be
fhewn.

Ihewn. For this liquid contains fo large a proportion of extraneous particles *, and thefe feem to have fo great an attraction for fire, that the fuperabundant fire is not fufficient to make them red hot. Perhaps there is not more fire feparated from the air in a given time by oil, than by fpirit of wine: but as oil contains a much lefs proportion of extraneous particles, the fame quantity of fire is fufficient to ignite the vapour of oil, though it cannot that of fpirit of wine.

If we fuppofe all other circumftances alike, the heat of the flame will be greateft in thofe bodies which, in a given time, faturate the greateft quantity of air, or generate the greateft quantity of fuperabundant fire. Perhaps lamp-oil has as great a proportion of phlogifton as zinc; yet the flame of zinc, if I am rightly informed, is by much the hotteft, and therefore the decompofition in the metal proceeds more rapidly than in the oil.

I USED formerly to think that thofe bodies would require the greateft heat to begin

* Water.

their

their combuſtion, which attraɗed the phlogiſton moſt ſtrongly; heat weakening the attraction between phlogiſton and earth. But that
this is not always the caſe appears from hence,
that if this rule held good, thoſe different bodies would ſuffer equal decompoſitions in equal
times, which does not agree with the laſt paragraph. In chymiſtry alſo, we find that the
phlogiſton may be transferred from the earth
of charcoal, to the calx of zinc; from thence
to the phoſphoric acid; from the phoſphoric
acid to the vitriolic; and from thence again
to the nitrous acid, or air. Yet charcoal burns
with more difficulty than ſulphur; and zinc
does not begin to flame but with an heat vaſtly
ſuperior to that which kindles phoſphorus.
Chymiſtry, however, furniſhes us with ſome-
thing analagous to this: thus all the acids, except the phoſphoric, attraɗ fixed alcali preferably to calcareous earth; but that acid attraɗs
the earth preferably to the alcali. The vitriolic acid attraɗs fixed alcali more than the
nitrous; but the nitrous acid attraɗs phlogiſton
ſtronger than the vitriolic. But one reaſon of
the greater difficulty in inflaming zinc than
                                    phoſphorus

phofphorus * may be, that the former contains
extraneous particles which fhield and defend
the phlogifton from the action of the air : thus
water defends phlogifton from the action of the
vitriolic acid, and fulphur cannot be formed
till by heat, or otherwife, the water is diffipated †. For that the difficulty of the combuftion
of zinc does not proceed from its fixity, or difficulty of being raifed into vapour alone, appears from hence, that the combuftion of lead
is effected without fuch vapour, or merely by
calcination. But when, by the action of heat,
the phlogifton of the zinc is rendered combi-
nable with air, the decompofition proceeds
rapidly indeed! by reafon of the weak attraction of the phlogifton for the calx, or of the
great eafe with which the air now attracts the
phlogifton from that body.

THE heat attending nitrous combuftions
may be underftood by referring to what was
faid concerning them in the fixth fection. A
certain degree of extraneous heat muft be ap-

---

* See alfo cafe vii. fection VIII.

† Fixable, and phlogifticated air may, for a like reafon, be
analogous to fulphur, and the volatile vitriolic acid.

plied

plied in order to enable the phlogifton and air of the nitre to combine, and fire is feparated from the air by the phlogifton, in the fame manner as from the air of the atmofphere in common combuftion.

BEFORE oils can be fet on fire by the nitrous acid a fufficient degree of heat muft be generated by their *chymical mixture** in order to enable the phlogifton and air to unite; and then the combuftion and confequent explofion happen as in ordinary nitrous combuftion.

THE generation of fixable air is a confequence of combuftion; and as this is generated in fermentation, in refpiration, and certain other proceffes †, there muft have been a combuftion. Thefe combuftions take place in the ordinary heat of the air, and are, properly fpeaking, *fpontaneous calcinations*. The heat is lefs than in inflammation, becaufe of the number of extraneous particles among which it is

---

* Hence the ufe of oil of vitriol. The like reafoning may be applied to the combuftion of pyrophyrus by common air.

† This is to be underftood in cafes where fixed air is actually generated, not where it pre-exifted and was only expelled as in effervefcent mixtures, &c.

fhared :

fhared : and the light of combuftion * is not
vifible by reafon that it is ftifled by thefe particles, and alfo becaufe it is too rare. Thus
fixed air may be produced by heating fulphur
in a clofe veffel ; and by repeating the procefs,
the whole fulphur might, I imagine, in time,
be decompofed as effectually as by inflammation, and yet by reafon of its rarity, the light
fhall not be vifible. Thus alfo coal, iron, liver
of fulphur, and other fubftances expofed to air
lofe their phlogifton in time ; and yet by reafon of the rarity of the light, and the flownefs
with which the decompofition proceeds, neither
light nor heat are fenfible.

By confidering the degree of affinity which
phlogifton has for the particles of bodies
with which it is combined, the volatility or
fixity of thefe particles, and the nature and
quantity of the extraneous particles which enter into the compofition of different bodies, we
may have the reafons why fome bodies inflame
with lefs heat than others; why they burn more
or lefs rapidly ; why the combuftion of fome
bodies cannot be effected without a continual

* Vide fection IX.

application

application of extraneous heat, and the like.
Hence alſo the difference between inflammation and incineration or calcination. The other
phenomena of the ſecond ſection will eaſily be
underſtood from hence, and from what has been
already ſaid.

WHETHER the hypotheſis on which I have
proceeded be true or falſe, experience muſt determine; thoſe who have leiſure and conve-
nience would do well to proſecute the Inquiry.
The Reader, on conſidering the importance of
the ſubject *, will excuſe me for detaining him
ſo long with conjectures; and the next ſection,
I hope, will make him ſome amends. I have
only been able to gueſs at the theory of the
heat of combuſtion; that of the light, at leaſt
the obſervational part of it, I think I can ven-
ture to offer as certainty.

* Vide ſections X. XI. and XII.

SECT.

SECTION IX.

*Of the Light and Colours which arise on the Ig-
nition, and Combustion of Bodies.*

SOME ingenious philosophers have of late
found that the calxes of certain metals may
be reduced by means of *light ;* and that other
phlogistic processes may be performed by it :
they have therefore imagined, that light is
phlogiston. The following considerations will
perhaps in some measure clear up this matter.

OBSERVATION I. When an incombustible
body is heated to a certain degree it emits light;
and the light increases as the heat becomes
greater.

OBS. II. The light which is first emitted is
of a reddish colour, so much so indeed that the
body is said to be *red-hot.*

L   OBS.

OBS. III. As the heat increafes, the colour verges more towards orange and yellow diluted with white; and when the heat is very intenfe, the colour becomes fo white that the body is faid to be *white-hot.*

OBS. IV. But if the heat be ever fo much increafed; yet, if there be no combuftion, the colour is never found to vary from the white towards blue, purple, or violet.

IT is known that light confifts of rays varioufly refrangible, and that this arifeth from the different fizes of their particles, thefe rays being moft eafily refracted whofe particles are the fmalleft; when thefe various particles of light are combined with a body, the leffer ones will be attracted and held moft powerfully, and the larger ones leaft, for the fame reafons that the rays compofed of them are differently refracted by that body. When air is combined with a fubftance, the application of a proper degree of heat will feparate it therefrom, and caufe it to fly off in its elaftic ftate. In like manner, when the particles of light are combined with a body, and heat be applied, thefe particles

will

will be diflodged, and expelled from the body
by the action of the fire, and as they have a
polar virtue (as appears by the double refrac-
tion of ifland chryftal) they will take the rectilinear difpofition, and conftitute *rays of light*.
But be this as it may, thofe particles which are
largeft, and which therefore are lefs forcibly
retained by the body, will begin to be diflodged
with a lefs degree of heat than the fmall ones,
which the body retains more powerfully; and
as thefe are the particles which conftitute the
red-making rays, the body muft appear red;
this red, however, will not be perfect, becaufe
fome of the other particles alfo will be expelled,
though in lefs quantity than is fufficient to compofe a white: as the heat increafes, the orange,
yellow, and other particles will be expelled in
more equal proportion, and therefore the colour
will verge from the red towards white, fo that
when the heat becomes fufficiently intenfe,
they will all be expelled alike, and the body
appear perfectly white. Thus in diftillaion,
when liquids of different volatility are contained in the alembic, if the fire be gentle, the
moft volatile will come over more pure; and

its purity will be lefs as the heat is more augmented.

But in thofe bodies which are combuftible,
we are prefented with very different phenomena;
if copper, for example, be heated, it will firft
fhine with a red heat, which will afterwards be
whiter according as the heat is increafed, as
fhewn above; but if it be made to flame, the
colour emitted will be green. Zinc may, in
like manner, be heated red-hot, and the red
colour will become whiter as the heat increafes.
But if it be made to flame, the colour is intenfely white. The flame of fulphur is blue,
of tallow, yellow, of lamp-oil, orange; and there
are hardly two bodies whofe flames are exactly
alike. Incombuftible bodies, therefore, when
ignited, and alfo thofe combuftible ones which
ignite before they flame, emit light in the following order; red, orange, yellow, green, blue,
indigo, violet. But this is not the cafe in combuftion.

Obs. V. The fubftance which burns with
the leaft heat of any that we know is phofphorus;

rus; the colour of its weak flame is a violet-
blue, if I am rightly informed. Sulphur burns
with an heat lefs than that of ignition, and its
colour (efpecially when the flame is gentle) is
blue; alcohol burns with an heat greater than
fulphur, yet below that of ignition, if properly
managed, and the colour emitted is blue,
though lefs fo than fulphur.

From hence it appears, that as in ignition
the red-making rays are moft copioufly emitted
(and at firft almoft entirely) ; thofe which are
emitted moft copioufly in combuftion are, on
the contrary, the violet. To account for this
difference, the Reader is requefted to attend to
the following reafoning.

It was fhewn in the foregoing fections, that
heat is neceffary to enable the air to attract
the phlogifton from bodies ; and that the fire
feparated from the air by the phlogifton, ferves
afterwards inftead of extraneous fire to keep
up the heat of the body, and enable the air to
continue the decompofition. The fire expelled
is fo copious as even to heat the body more

L 3   than

than is fufficient to enable the air to attract the phlogifton, as is evident by the flame of a combuftible body fetting fire to a body which requires a greater heat than it to begin the combuftion; the intenfenefs of the heat, therefore, and the violent attraction of the air, will diflodge the phlogifton from the body fafter than the air (efpecially when it begins to be fatiated) can readily combine with it; and thofe particles which are not immediately combined, attract large atmofpheres of æther, which render them incapable of combination with air, and therefore they are driven off in the form of *light*.

Now, as in ignition, bodies retain the blue light moft powerfully, and part chiefly with the red, and other particles which compofe the lefs refrangible rays; fo in combuftion, the air moft eafily difengages and attracts the larger particles, with which being firft nigh faturated, the fmaller ones remain behind as the fuperabundant particles above fpoken of; and which, by acquiring ætherial atmofpheres, become particles (and are driven off in the form) of light.

The

The flame therefore muſt appear of a colour oh the violet ſide of white, as we find to be the caſe *,

BODIES which ſhine by ignition can, for rea-ſons juſt given, advance in colour only from red, through a dilute orange and yellow, to white, and can never paſs from that white to green, blue, and violet; ſo neither can the light of combuſtion paſs on to yellow, orange, or red; but yet we find that ſome flames are tinctured with theſe colours. Thus the flame of a candle is yellow, that of an oil-lamp orange, and of wood red.

BUT it muſt be obſerved that the above rule holds good only in thoſe flames whoſe heat is below ignition. When the heat is in-tenſe, the particles which compoſe the vapour

---

* The ſmaller particles will alſo attract æther faſter than the large ones. The reaſon that light does not thus appear when the vitriolic acid &c. takes phlogiſton from bodies ſeems to be, that by reaſon that the particles have leſs atmoſpheres of fire, they are nearer in contact with the phlogiſton, and therefore by attracting it prevents its eſcape; or the light expelled may be too rare to be viſible for the ſame reaſon.

L 4   are

are ignited, and the light which proceeds from
them is mixed with the light of the combuftion; but the light of the ignited vapour is
emitted in a contrary order to that of the combuftion, the latter beginning at violet, the former at red * ; and therefore the mixt colour
of the flame will be varied according to the
degree of heat, the proportion of the light of
the combuftion to that of the ignition (which
in fome cafes will depend on the nature and
proportion of extraneous particles † in the vapour), and to other circumftances. Thus, the
flame of wood feems to be in the firft degree
of ignition, or *red-hot*, and the proportion of
the light, to that of the combuftion, is fo great

---

* As a farther proof of the difference obferved, thofe flames
which are ignited are opaque; but thofe that are not ignited
are tranfparent. The flames of oil, and of alcohol, properly
managed, will fhew this to advantage.

† How thefe affect the heat of the flame (on which its ignition depends) may be gathered perhaps from the laft fection;
and as the ignition is lefs, its light is redder, as was fhewn
above. Different bodies may alfo, perhaps, have different
proportions of the blue light in their compofition. Some flames
feem likewife not to have any, or but very few, extraneous
particles, and therefore are ftill blue, though their heat be
great.

that

that its colour is predominant; the flame of oil, though lefs red, does not feem to be much hotter than that of wood, but the light of the combuftion is in greater proportion, fo as to dilute the red to an orange. The fame may be obferved of the flame of a candle which is of a dilute yellow; and in the like view the colours of flames of other bodies may be confidered, fome of which I have arranged in the following table.

TABLE

TABLE OF THE COLOURS OF FLAMES.

Bodies arranged according to the degrees of heat neceffary to begin their combuftion.	Light of the combuftion.	Light of the Ignition.	Proportion of the lights of C, and I.	Colour of the flame.
The weak flame of phofphorus	Violet-Blue	None	All C	Violet-Blue
Sulphur	Blue	None	All C	Blue
Alcohol	Greener Blue	None	All C	Dit, a little greener
Small wood	Ditto?	Red-white	I, moft	Reddifh white
A pitch torch	Ditto?	Ditto?	Ditto	Ditto
Lamp-oil	Ditto?	Ditto?	Ditto, lefs	Orange-white
Tallow	Ditto	Ditto?	Ditto, ftill lefs	Yellow-white
Camphire	Ditto	Ditto?	Ditto, ftill lefs	Ditto more white
Nitre and coal	Ditto?	Yellow-white	I, moft?	Ditto ftill more
Copper	Ditto?	Ditto?	C, moft?	Green-white
Iron	Ditto?	Whitifh	I, moft	White
Zinc	Ditto?	Whitifh	Ditto, lefs	Whiter

Others

Others might have been added; but even thofe which I have given are very incorrect, being fet down only by guefs; and the table is offered merely as a fketch of the fubject, to be profecuted by thofe who have proper inftruments, and other conveniencies.

It is to be noted, however, that different parts of the fame flame are unlike in colour: thus, the bottom of the flame of a candle or oil-lamp is blue; it grows lefs blue by degrees till it ends in a yellowifh white; but this white, towards the top, verges towards orange, and ftill further, towards red (efpecially when the flame is advantageoufly difpofed), till it ends in unignited vapour or fmoke.

To underftand the reafon of this it muft be obferved, that at the bottom, where the decompofition begins, the light emitted is only that of the *combuftion;* for it takes up fome little time to ignite the particles; and therefore they do not begin to emit the light of ignition till they have afcended fome way up in the vapour. But when that takes place, the blue colour of the combuftion begins to be changed, till at laft

the

the mixed colour in the middle is of a yellowifh white. But the light of the combuftion
being lefs and lefs towards the top, till perhaps
it quite ceafes, and the furrounding air cooling
the ignited particles into a lefs white heat *,
the colour of the flame towards that part is
more red, till at laft the particles lofe their
fhining heat, and pafs off in the form of unignited vapour. When the wick is long, and alfo
ignited, the latter phenomena are more confpi-
cuous; for the red light of the wick, and of
the particles that efcape from it, being mixed
with that of the flame, tinctures it, efpecially
at the point, more highly with red. In day-
light, or fun-fhine, the latter phenomena appear
to ftill greater advantage, the weak light of the
combuftion being then lefs capable of inter-
rupting that of the ignition.

THE colour of the light of the combuftion
of bodies may be known by obferving the bottom, where it is as yet unaltered by that of
the ignition, for reafons given above. Thus,
if you faften a piece of camphire on a wire, and
inflame it, holding it up in the air, you will fee

* Hence the Conic form of the flame.

a blue

a blue light at the bottom. The light of the combuftion in all bodies muft be more or lefs blue, becaufe mixtures of the moft refrangible rays produce only various fhades of that colour *.

In the above table only the middle part of the flame is eonfidered, where the light is compound, as in tallow ; and only in the weakeft ftate, where it is fimple, as in alcohol. But if a quantity of alcohol be burnt, fo that the flame rifes high, the particles will be ignited. And if we examine the upper part of the flame, and compare the colour with what has been faid, we fhall find this to be the cafe. The other phenomena of the lights of ignition, and combuftion, either feparate, or conjoined, may perhaps be underftood by profecuting the principles above laid down.

Corollary I. Bodies retain a confiderable

* Perhaps even the ftrongeft ignition that we can caufe by our fires does not yield a perfectly white heat. The flame of zinc, however, is intenfely white, if I am righfly informed. The light of the ignition is, by that of the combuftion, diluted to a perfect white.

quantity of particles of light in their pores, or otherwise. Thefe particles are diflodged and expelled from thofe bodies by a proper degree of heat, and the largeft particles moft eafily, by reafon that they are lefs forcibly retained. Hence the light of ignition.

Cor. II. Phlogifton *combined* with bodies cannot be expelled by heat alone, though light can ; thus charcoal, heated in a clofe veffel, though it may be made to emit light, yet is not found to part with its combined phlogifton; yet the light of the combuftion is this very phlogifton fet at liberty by the combined action of heat, and the attraction of the air. Vide fections VII. and IX.

Cor. III. Phlogifton therefore is light in a ftate of combination with bodies, forming a conftituent or effential part of them. Light is phlogifton in an elaftic ftate exifting in their pores. As this laft is lefs attracted by bodies, they fhine with a lefs heat : thus electricity and certain phofphori, fhine with the ufual heat of the atmofphere ; and fome of the latter, if expofed to any particular fort of the fun's

rays,

rays, expel them again in the fhade; that is, the fame colour which was forced into the body is afterwards emitted by it, as its attraction for fire, which was diminifhed by the action of the light, returns.

THAT the light of thefe phofphori is what I call *the light of ignition* appears by its colour. In fome cafes of electricity, however, a blue light is obferved; but this does not happen unlefs a *real combuftion* takes place, and that this does fometimes obtain is obvious by an experiment of Dr. Prieftley, who made fixable air with this fluid. I had drawn up a theory of electricity, and intended that it fhould have followed this fection; but found, after I had finifhed it, that it would fwell the volume to a much greater fize than was intended. The electric fluid appeared to me to be *phlogifton combined with earth, already more intimately combined with a confiderable portion of that principle;* for air takes it from that earth, as appears by the above experiment; neither can pure phlogifton combine directly with air, for a reafon to be met with in the 7th cafe of the VIIth. fection. That the earth is of this nature appears

appears by the fulphureous or phofphoric fmell, and by its changing blue infufions red. The quantity of this earth is not fufficient to render the phlogifton coherent ; but the difference between pure phlogifton and electricity, feems to be fomewhat the fame as between pure and fixable air. From cafe 13th, fection VIIth. I had inferred *that as by friction heat is generated, it argues that by friction the attraction between the phlogifton and earth is increafed.* Hence when glafs is rubbed by the hand, their attractions for phlogifton are both increafed, but that of the glafs (being the ftrongeft electric) moft. The glafs therefore will attract it from the hand, and the hand from thofe conductors which are in contact with it. Yet not the phlogifton, combined in a coherent form in bodies, flows, in this cafe, to reftore· the equilibrium or common temperature, but only that which exifts in the pores in an elaftic ftate. Pure phlogifton will flow as well as the other ; hence the *electrical light of ignition.* By the friction fome of this phlogifton will alfo, perhaps, be converted into electricity, being attracted by the excited effluvia of the hand: and I had gone through all the
principal

principal phenomena of electricity. If this
sketch of the subject, and the present work be
approved, I may hereafter publish the original
essay, together with other papers on different
subjects.

In combustion, and some other chymical
mixtures, perhaps a small part of the heat may
be occasioned by the friction or percussion of
some of the particles, though much less than is
at present believed : but bodies are probably
heated by light *, and by electricity, in great
measure by this means ; and entirely by it and
the communication of phlogiston, except in
cafes where actual combustion is caufed.

By the second corollary it appears, that phlogiston, when disengaged from bodies with
which it was combined in a coherent state,
assumes the form of light ; and it was shewn
before, that fire when disengaged is the caufe
of heat. I had formerly run the analogy between these principles farther, and imagined
that fire did not only exist in bodies after
the manner already described, but that it alfo

* Hence opaque and denfe bodies are moft heated by light.

M   combined

combined with them in a coherent form, was disengaged by phlogiston, &c. and then assumed its elastic state: and also, that it was transferrable from one body to another in a fixed state, in the same manner as phlogiston *. But I could not find means to satisfy myself of the truth of these propositions, and the mode of its existence, which I have before supposed seems to agree with the phenomena of heat and cold as exhibited by the sense, and the thermometer; but the truth remains to be cleared up by experiments. In the mean time the hypothesis that " phlogiston weakens the attraction of earth for fire, according to the force of their combination," and that " the force of their combination may be intended, or remitted," seem sufficient to account for the phenomena, whatever

---

* For example: I argued that causticity depended on fixed fire. That a fixed alcali being applied in its mild state to quicklime, the lime combined with the fixed air, and the alcali with the fire. That water expelled fire from the vitriolic acid, quicklime, &c. by means of a superior affinity, as fixed air is expelled from mild alcalis by acids. That spirit of vitriol added to caustic alcali, the acid joined with the alcali, and the fire with the water, the fire being more than sufficient to saturate the water, &c. &c. The like of air. But these things may equally obtain on either supposition, and what respects causticity does not seem to be true.

be

be the modes in which fire and phlogiſton exiſt in bodies.

To the mode of its exiſtence which I have ſuppoſed, it may be objected, that if the particles of bodies have the atmoſpheres of fire deſcribed, the bodies which they compoſe ought to repel each other like particles of air; for the atmoſpheres of the particles extending beyond the body, will compoſe a repelling atmoſphere of fire about that body: it may be anſwered, that theſe repelling atmoſpheres are, naturally, balanced by means of electricity. When the latter is removed, the action of the former becomes ſenſible; for two bodies negatively electrified repel each other; or, if the equilibrium be deſtroyed in a contrary way, by the attraction in conſequence of excitation, or by accumulation of electricity, theſe repelling atmoſpheres are equally left at liberty to manifeſt their action; and the electricity may even conſpire with it, if great; hence the mutual repulſion of two bodies electrified poſitively *.

For

---

* Electrical repulſion, whether *plus* or *minus*, ſeems to depend on the atmoſpheres of fire, as above. But electrical at-

traction,

For a reafon given above, I cannot now enlarge on this fubject, and therefore fhall only add, that if the repulfion of the particles of air is not diminifhed by combuftion fo much as might be expected, the caufe may perhaps partly be difcovered from hence; and alfo by confidering that particles of air, &c. may probably be only bodies made up of other particles, thefe again of others, &c.; and that the atmofphere of a whole particle is made up of thefe portions of the atmofpheres of its elements which extend beyond the whole particle. Hence the more the particle is condenfed by cold, or by combination with phlogifton, more fire in proportion will come into the atmofphere from its pores; the fire fo expelled will alfo be more expanded. And contrariwife when the whole particle is expanded by heat, or the lofs of phlogif-

---

traction, on the violence with which bodies attract electricity. Thus, a non-electric being properly prefented to a body *minus*, the latter attracts the electricity in the former (and with it the body itfelf) fo violently as to exceed the mutual repulfion by their atmofpheres of fire. When the electricity in the two bodies becomes in equilibrio, if the quantities be natural, the attraction and repulfion ceafe; but if they be ftill either *plus* or *minus*, the bodies repel each other by means of their atmofpheres of fire, as before.

ton,

ton, the repulfion of the whole particle may depend only on the fire in the atmofphere thereof.
But if we fuppofe part of the fire to have been
fixed *, the folution is perhaps ftill more eafy.

WHEN air is applied to zinc, and to moft
other combuftible bodies, under proper circumftances, it deprives thofe bodies of a certain
portion of their phlogifton ; but another portion remains behind, of which air cannot de-
prive them. The fixed particles of bodies,
therefore, or what the chymifts call earth, have
ftronger attractions for phlogifton than even
air, and therefore are *originally* † more pure
earth in my fenfe of the word. The phlogifton which remains, and which cannot perhaps
be taken from them by art, is fufficient even
to keep them coherent. If they were deprived
of that extra phlogifton, they would therefore
form a fluid as much more elaftic than our air,

---

* May not the very condenfed fire next to the furface of
an ultimate particle be faid to be *fixed?* Does fire combine
with bodies in any other manner than that above defcribed ?

† Vide cafe xvii. fect. VII,

as

as their attraction for fire would be ftronger, if the foregoing reafoning be true.

IN phlogiftic atmofpheres, the fmalleft particles, being moft attracted, may immediately furround the particle; thofe which are larger next above, and fo on to the largeft, which may compofe the external part of the atmofphere. Hence may be another reafon why air in combuftion firft, and moft eafily attracts the larger particles, as obferved before. Hence alfo perhaps the reafon that the fpheres of cohefion are fo limited. The like atmofpheres may be fuppofed of the elements of a proximate particle, of the elements again of thefe, &c.; the latter will retain their phlogifton more powerfully than the former, as being lefs; and their cohefion will, for the fame reafon, be ftronger. This, were it true, might enable us to account for what was obferved in the laft paragraph; alfo for the heat generated by oil of vitriol and water, &c. and (together with what was obferved of the manner in which fire is combined) gives us fome idea of the internal ftructure of bodies.

ASTRONOMERS

ASTRONOMERS freeze at the thought of the planet Saturn, and entertain a contrary fentiment with refpect to Mercury ; but if the proportions of fire, &c. in the different planets be properly adjufted, as their denfities feem to fhew, it may not be fo hot in Mercury, nor fo cold in Saturn as is at prefent believed. Whether the like reafoning may be applied to the fun and fixed ftars ? (See what was faid above concerning phofphori.)

M 4   SECT.

SECTION X.

Of Refpiration, and Animal Heat.

I FIND by fome extracts from Dr. Prieftley's publications, that that great philofopher has demonftrated *that the ufe of refpiration is to carry off the phlogifton which the blood acquires during its circulation in the body.*

IT feems to apppear that there is a very clofe analogy between refpiration and combuftion; and this has been an ancient obfervation. Dr. Willis of the laft century, treating on the heat of the blood, has this paffage. " Though it feems an hard faying that *the blood is accended,* yet feeing we can attribute its incalefcence to no other caufe, why fhould we not impute it to this? efpecially feeing *the proper paffions of fire and flame* agree to the *life of the blood.*

" FOR the chief and moft effential requifites to continue a flame are thefe three: 1*ft.* That a free

a free and continual accefs of air be granted
to it as foon as it is kindled. 2dly, That it en-
joy a conftant fulphureous *pabulum* or fewel.
And 3dly, That as well its fuliginous as thicker
recrements be continually amanded from it :
and feeing thefe agree to the *vital flame*, as
well as to an elementary, it feems very rational
to affirm that *life itfelf is a kind of flame.*"

THIS learned and very ingenious phyfician
faw plainly that there was an analogy between
combuftion and refpiration ; and between the
heat of flame, and that of the blood. But for
want of proper difcoveries concerning the nature
of combuftion, &c. his ideas were more confufed and obfcure. A little more light may perhaps be thrown on the fubject in the courfe of
this fection, and yet fucceeding authors, who
pufh their inquiries farther, will make a fimilar obfervation on what I have done,

ACCORDING to the difcovery of the great
philofopher above mentioned, the blood which
is brought to the lungs from the body, contains a greater quantity of phlogifton than that
which goes from the lungs into the body. And
the

the air takes this phlogifton from it in its paffage through the lungs.

IT appears, by what was faid in the Vth.
fection, that when phlogifton end air combine,
heat is generated. And if the conjectures in
fection VII. can be depended on, the heat arofe
from hence, that the phlogifton by combining
with the air weakened its attraction for fire,
which therefore gravitated towards, or was attracted by, the bodies around, till an equilibrium, with regard to the attracting powers of
the refpective bodies, again took place. When
therefore the particles of air combine with the
phlogifton which they attract from the blood
in the lungs, heat will in like manner be generated : that is, the attraction of the particles of
air for fire will be diminifhed as in combuftion.
The blood which is carried to the heart therefore will be hotter than that which is brought
to the lungs ; and hence one caufe of the heat
of the blood.

BUT when the particles of blood loft their
phlogifton to the air, their attraction for fire,
which by cafe 7th. fection VIIth. was weakened

by

by the phlogifton, will, now they are deprived
of that principle, be again increafed. A part
of the fuperabundant fire, therefore, will be attracted by thofe particles of blood; and they
will be carried in this ftate into the body.

It is known to philologifts, that if the nerves
which ferve any particular part be deftroyed,
that part will be colder than before, notwithftanding that the blood circulates through it as
ufual. Now as the blood, when it entered
that part was already hot, and as, before the
nerves were deftroyed, the heat of the blood was
fupported in its paffage through that part, and
on the contrary, when the nerves were deftroyed, the blood was cooled in its paffage through
it; it follows *that the heat of the blood is fupported in its paffage through a part, by means of
the nerves by which it is ferved.*

The heat in combuftion feemed to arife
from hence; that when the particles of air
combined with phlogifton, their attractions for
fire were diminifhed: may not the fame reafoning be applied to the blood? as the blood
in the veins of the body is found to contain
more

more phlogifton than that in the arteries; and
as the heat of the blood in the body appears to
depend on the nerves, may we not argue in
the manner following ?

THE blood in its paffage from the arteries
to the veins has phlogifton imparted to it either
immediately or mediately, by the nerves. But
each particle of blood thus combined with
phlogifton will have its attraction for fire leffened, analogous to what happens to a particle
of air in combuftion. Heat therefore will follow for the fame reafon that it follows on the
combination of phlogifton with air in combuftion, only in a lefs degree. The particles of
blood thus phlogifticated, and rendered unfit
for the further purpofe of caufing heat, pafs on
with the circulation, and frefh ones fucceed.
When the phlogifticated particles arrive at the
lungs, they are decompofed by the air which
attracts their phlogifton, and from which the
particles of blood, in return, take a quantity of
fire, fo that they are again rendered fit for the
purpofe which has been defcribed. And this
feems to me to be the manner in which the
blood becomes hot.

                                    S E C T.

SECTION XI.

Of the vital and other motions of the Body.

IF some purpose of the last importance to the animal had not been designed by *respiration*, the all-wise Author of nature would, certainly, not have rendered life so dependant on that process as to be incapable of existing, even a few minutes, without it.

BY the last section it appeared probable, that the heat of the blood depends on the *nerves;* or that the phlogiston which the arterial blood acquires in its passage to the veins, is communicated to it by those organs.

ALL the vital motion or functions of an animal body are performed by means of the nerves ; and all those functions may be reduced

to

to the *contraction of the moving fibres* *. I would fay, therefore, that *for a nerve to caufe the action of a fibre, it is neceffary that the nerve fhould impart phlogifton, either immediately or mediately, to the blood flowing through, or by that fibre.*

THAT there is a connection between the action of the fibres, and the phlogiftication of the blood, appears, I think, by the following confiderations. I. The heat of the blood depends on the nerves, as appears by the laft feetion. II. According as more of the voluntary mufcles act, or as their action is ftronger, more blood is phlogifticated in a given time; for the heat generated is greater, and the refpirations are quicker: and III. The motion of the blood through a mufcle is known to be as neceffary to its action as the nerve; for if the artery be tied, the mufcle becomes paralytic as effectually as if the ligature had been made on the nerve.

THE action of the moving fibres may be divided into voluntary, and involuntary : fome

* Senfation is not here confidered.

fibres ferve for involuntary action alone; others for voluntary; but thofe mufcles which are for the voluntary motions of the body, are continually exerting involuntary action. The contractions of the arteries, the veins, and other veffels of the body for the purpofe of circulating the fluids, &c. are performed by means of moving fibres. The mufcles, membranes, coats of veffels, &c. are made up of fuch fibres; there is no fenfible part of the body but what abounds with them; all thefe are continually exerting involuntary, and moft of them in walking, voluntary actions, neceffary to the life and well-being of the animal. Now, as there feems to be a mutual dependence between thefe actions, and the phlogiftication of the blood, as the number of particles of blood is not infinite, but on the contrary, only fuch a quantity can be admitted into the ftructure of the animal fabric as is fufficient to balance the action of the folids, if there was no contrivance for dephlogifticating the blood, the whole mafs would foon be rendered unfit for the purpofe juft defcribed, as well as of communicating heat, and death would prefently enfue. Nature has there-

fore

fore provided the animal with lungs; the blood,
phlogifticated as already related, is conveyed to
that organ; the air in infpiration reftores it to its
original purity by taking from it its phlogifton,
and furnifhing it in return with fire, and thus
renders it again fit for the purpofes of animal
motion and heat. In proportion therefore as the
fum of the whole action of the fibres of an animal is greater, that is, in proportion as a greater
quantity of blood is phlogifticated in a given
time, the motion of the blood ought to be increafed, and the infpirations of air more fre-
quent, in order that the reftauration of the
blood to its former purity, may keep pace with
its phlogiftication in the body.

Now, as life depends on the action of the
fibres, as above, as there is a neceffary connection or dependence between the action of thefe
fibres and the phlogiftication of the blood; and
as from the great number of moving fibres in
the body in continual action, and the fmall
quantity of blood, the latter will be prefently
phlogifticated, we have an idea of the very
great importance of refpiration, and the abfolute

lute neceſſity of it to the continuance of life, as we find by experience to be the caſe; neither the *heat of the blood*, nor even *the vital motions of the ſyſtem* being capable of exiſting long without it.

N   S E C T.

SECTION XII.

Of the Action of the Fibres, or muscular Motion.

WHAT has been said in the two laft fections may be allowed, perhaps, to be in fome degree probable. I fhall give no opinion with regard to what follows.

THE idea of mufcular motion, which I had formed to myfelf many years ago, was, that *by the influence of the nerve, the particles which compofe a mufcular fibre had their attractive forces increafed, fo that they were drawn nearer together, but that as foon as that influence ceafed, the increafe of attraction vanifhed, and the particles receded to their previous diftance from each other.* I had contented myfelf with a theory in the abftract; but Dr. Prieftley's admirable difcovery will, perhaps, enable me to affign the phyfical caufe of this contraction.

IT

IT appears probable to me, after an attentive confideration of the fubject, that the matter or fluid contained in the nerves which
ferve for motion, is *the phlogifton, combined in
a coherent form with an earth already more intimately blended with a confiderable quantity of
that principle ; fo that their combination is but
weak* *. Thofe who have read the feventh fection carefully will comprehend my meaning by
this definition, and therefore I need not com-
ment upon it. This matter does not feem to
be derived from the nerve into the fibre of itfelf, or by propulfion, like the blood, for if the
nerve be tied, it does not fwell between the ligature and the brain. The matter of the voluntary nerves is, I think, only driven down by
the will †. That of the involuntary ones is obtained by means of the pulfe of the arterial
blood, and other ftimuli in the body, by *irrita-*

---

* Some phenomena feem to fhew that the latter ingredient only is fecreted by the brain, and that the former
bibed from the ftomach, &c. That the nervous fluid is not
the electrical matter, as fome have fuppofed, is plain from its
not combining with the blood in the manner the phlogifton
in queftion is found to do.

† Pain is a ftimulus to thefe nerves: but then it is by its
action on the fenfory, &c.

tion,

tion, or *reflux*. Hence perhaps one reason *
why the blood does not move in a smooth un-
interrupted course in the arteries, but by pulses;
and hence also the reason that it does not flow
by pulses in the veins, the fibrous mechanism
terminating where the veins commence, so that
there is no further occasion for it. The reason
of all this seems to be, *that such a quantity of*
matter only may be occasionally derived from the
nerves, as may be necessary for the purposes of
the animal economy, &c. which, therefore, is
left to be regulated by the will, by the pulse of
the blood (the force of which depends on muscular action), by heat, and other stimuli.

WHEN by the pulse of the blood, the influence of the will, &c. a portion of this matter
is derived from a nerve into a fibre, it seems
to me that the particles of which the fibre is
composed, having a greater attraction for the
phlogiston, than the earth has with which it is
already combined, take the phlogiston from that
matter, and thereby have their force of cohe-

---

* The other reason seems to be that the fibres may be put
into vibrations, by means of which the effect mentioned is
also probably produced. Of these vibrations I may speak
more at large in future.

fion increafed; the fibre, therefore, will con-
tract: but the particles of blood flowing by the
fibre, and having a ftill greater attraction for
phlogifton, takes it immediately from the fibre,
which therefore is again relaxed. Hence, as
the contraction of the fibre is but momentary,
if its contraction be required to be continued
for a given time, there muft be a continual de-
rivation of matter into it from the nerve during that time *.

It may be afked, that if mufcular motion be
performed by means of phlogifton caufing a
temporary increafe of attraction or cohefion in
the particles of the fibres †, why this indirect
method

---

* I endeavoured to account for mufcular motion by the
phlogiftication and confequent contraction of the blood only,
and alfo by the æther difengaged from the phlogifton by
the ftronger combination. (Vide fection VII.) But neither
of thefe by any means anfwer to the phenomena. It may be
obferved that, probably, only the craffamentum, or its coagu-
lable lymph, attracts the phlogifton from the fibres. That
phlogifton when combined brings particles nearer to each other
is plain by its effect on air, metallic calxes, &c.

† If an artery be compreffed, a fenfation of warmth is perceived in the part which it ferves; but as the blood returns,
cold is felt. The warmth arofe perhaps from the fibres being

phlogifticated,

method was adopted, and why the nerves were not furnifhed with it as a fluid, fo that it might have been derived from them immediately to the fibres ? It may be anfwered, that probably phlogifton cannot be managed thus *per fe* ; and if it could, yet the quantity which a nerve would contain would not perhaps be fufficient for a fingle contraction of a mufcle : whereas, by this contrivance, a nerve can contain a fufficient quantity to laft a long time. But there is, probably, ftill another reafon ; it has been an opinion of long ftanding that the parts of the body are nourifhed either wholly, or chiefly, by the nerves; for a part rendered paralytic by dividing a nerve waftes, notwithftanding

phlogifticated, and the blood not being able to take the phlogifton from them, by which their attractions for fire continued diminifhed. But when the blood flowed again, and attracted the phlogifton from the fibres, cold muft have been the confequence, by the theory of combuftion before explained. Alfo, when the artery only is comprefled, the fibres feem to be more rigid or contracted than naturally. But when only the nerve is comprefled, the fibres feem, on the contrary, to be more relaxed. If the experiments which I have made on myfelf (of which thefe conclufions are the refult) can be depended on, they furnifh a kind of proof of the theory of mufcular motion above laid down. It is alfo known that a mufcle does not fwell when it contracts.

that

that the blood flows through it as ufual. If the foregoing conjecture be true, the nervous matter is a compound of phlogifton, and an highly phlogifticated earth ; and each of thefe ingredients may have their refpective ufes. The ufe of the phlogifton may be to caufe the contraction of the fibres, and the heat of the blood. that of the other ingredient (the phlogifticated earth) to nourifh the fibres, &c. not perhaps alone, but conjointly with the blood ; and hence the attraction of the fibres for phlogifton is between that of the nervous earth, and the blood. Hence alfo the gelatinous nature of the fibres. Hence people who ufe no exercife have their flefh more delicate and fat than thofe who labour hard, the nervous matter of the former not being fo liable to be carried off, but enters more into the compofition of the fibres *, and fome of it, perhaps, even in an undecompofed ftate. In the hands of a Pringle, or a Fothergill, thefe obfervations, and others which have been given, might, perhaps, be rendered of ufe in the practice of phyfic. It may be added, that as oil of vitriol cannot decompofe char.

---

* Hence the neceffity of reft, or fleep appears.

coal,

coal, fo blood may not be able to decompofe the nervous matter itfelf, though it fo readily takes its phlogifton from the fibres. The nervous medulla is not eafily combuftible by *air*, if I remember right, notwithftanding that it is fo readily decompofed by the moving fibres, if the foregoing conjectures be true.

IT may alfo be objected, that a longer time feems neceffary for this procefs than appears to be confiftent with the inftant contraction of a mufcle from the influence of the will. But not to mention how quick the tranfition of fo fubtile a principle as phlogifton may be effected, I could, I think, clear up this difficulty by quotations from papers on the fubject; but as that would lead me too far out of my way, I fhall only obferve, that the perception which we call *willing*, and which we ufually confider as the caufe of the action of a voluntary mufcle, is only an effect of the fame caufe in the fenfory, by which the contraction of the mufcle is brought about. And, to illuftrate the refult by a fimile, as when a man is fhooting at a mark, and we ftand near that mark at a diftance from the man, the fhots are heard to ftrike againft the pa-

per

per as foon as the report of the gun ; fo, for a
reafon fomewhat fimilar, the volition and the
action, feem to us to be in a manner cotempo-
rary. It may likewife be remarked that our
perceptions and actions are exceedingly flow,
when compared with the action of the more
fubtile principles of nature, as I may hereafter
explain, if the prefent work be approved.

A CERTAIN degree of heat, though neceffary to a particular fpecies of animals, is by no
means fo to the animal functions, or to animal life in general. Thus fifhes are as perfect
in thefe refpects as quadrupeds, though their
heat be much lefs. The heat is neceffary to
the liquifaction of the blood, and, perhaps, of
the nervous compound ; and alfo, to enable the
fibres the better to decompofe that compound,
and the blood again the fibres. It may hereafter be fhewn that it alfo probably affifts in the
vibrations of thofe fibres. The blood of fifhes
is fluid with a degree of heat in which that of
quadrupeds would be congealed. Animals
which require much heat to keep their blood,
&c. fufficiently fluid, decompofe a proportionally larger quantity of air, fo that their blood
may

may be more heated in the lungs, and alfo that its particles may carry a greater quantity of fire into the body to be extricated by the phlogifton from the nerves. Fifhes, whofe heat is required to be but little, decompofe a fmaller quantity of air, in an equal time than quadrupeds; and the air which is feparated from the water by their gills, and again purified by water, may be fufficient for that purpofe. Now as lefs fire is feparated from the blood of fifhes than from that of quadrupeds, it argues that lefs phlogifton is alfo imparted to the blood by their nerves: and this agrees with an ob-fervation of phyfiologifts, that *the fibres of cold animals are more irritable than thofe of hot ones.* The balance therefore is preferved; for as lefs *heat* is required to liquify their blood, &c. than in land animals, fo lefs *phlogifton* is neceffary to the contraction of their fibres.

It would be eafy to enlarge on fuch a fub-ject as this; but as I only offer what has been faid as fpeculation, and by way of hint to be profecuted by others, I fhall not *here* purfue the idea any farther. I will only add a wifh hat what I have offered, may not give occafion to the barbarity of making experiments on living animals. Were I even certain that this theory could be proved by making fuch experiments, I would not attempt them, as I do not think we are by any means warranted in putting animals to torture to gratify philofophical curiofity : I would not be underftood as fpeaking this from a principle of fuperftition; it is dictated by my own feelings : that man who has experienced in himfelf the extremity of *pain*, muft be fomething worfe than I can imagine to inflict it on animals, who are incapable by their natures of giving him caufe.

APPENDIX.

APPENDIX.

SINCE the foregoing work was finished, a book has been published on the subjects of animal heat, &c. by Dr. Leslie. The very favourable account which the Reviews gave of that work, and the deference which I paid to the judgment of the authors of them, made me at first doubt whether I had not proceeded entirely on wrong principles in my inquiries on the same subjects, and had resolved to withhold either the whole Essay on Combustion, or at least that part of it which treats of respiration, &c. from public view; but after considering the matter more attentively, and comparing the different performances, I thought I discovered reasons for imagining that my arguments were not so fallacious as I had at first concluded. I

was

was at length fo far fatisfied of this, that I drew
up a paper by way of refutation of Dr. Leflie's
theory, and intended to have fubjoined it to
my Effay ; but after the work was in the prefs,
another performance on the fame fubject was
put into my hands, written by Mr. Crawford *, which contained experiments totally
fubverfive of Dr. Leflie's principles, and there--
fore rendered my refutation, which was chiefly
fpeculative, needlefs.

WITH refpect to the latter performance, I
do not hefitate to pronounce it one of the beft
philofophical pieces that the prefent age has produced. I have no knowledge whatever of the
Author, any more than I have of Dr. Leflie, and
therefore can have no other motive for praifing
his work than a fenfe of its merit : I read it
with pleafure, not only on account of the new
and important points of philofophy which it
unfolds, but of the truly ingenious and philofophical manner in which he has treated his fubjects. We are here prefented with a fpecimen of
the true method of inveftigation in philofophical

* I had not the pleafure of feeing this excellent work till
fome time after it was publifhed.

matters,

matters, and with an example of that becoming
modefty which always accompanies real genius.

THIS juftice which I do to Mr. Crawford's
performance may be confidered as the more fin-
cere as I frankly acknowledge that the pleafure
with which I perufed it was not unaccompanied
with pain. I had been writing on the fame
fubject ; had advanced a ftep farther than
others had done, and was about to make my
difcovery public. Had I been fortunate enough
to have gone to the prefs before the ingenious
Author I am fpeaking of, a fmall degree of fame
would probably have been acquired by the
publication: but this Gentleman has outfoared
me; he has completed the fubject which I had
only begun, and proved by facts, propofitions
which I had only offered as conjectures. Un-
happily I have been fo circumftanced that it
has not been convenient for me to engage in a
courfe of experiments requifite to the complete inveftigation of a philofophical truth. I
could only argue from thofe facts which had
been publifhed by others, and from analogy ;
yet by the help of thefe I was enabled to difcover that combuftion is a truly chymical procefs,

cefs, and that it depends on the fuperior affinity or attraction between phlogiſton and air :
that by the combination of thefe, a degree of
heat is generated fufficient to account for the
heat and light of flame, and for the continuance
of the combuſtion after once begun. But the
origin of that heat I had only dreamt of ; and
even my dream I find did not wholly correfpond
with the truth : yet had I publifhed firſt, I fhould
perhaps have deprived our illuſtrious Author of
fome part of the glory which he has gained by
his excellent performance : I fhould at leaſt
have had the credit of difcovering the propofitions eſtablifhed in the third and three fubfe-
quent fections of the Effay, and perhaps of fur-
niſhing the *hint* of the remainder of the fubject.
but Mr. Crawford has fairly got the ſtart of
me.

THIS, however, is not the only inſtance of
two perfons unknown to, and ignorant of each
others purfuits happening to hit upon the fame
difcovery. That mine were made independant
of that Gentleman's, appears by a view of the
two performances ; for it will be feen that we
proceeded on a· very different rationale, and
that

that we arrived at the fame conclufion by direct oppofite roads. My learned and worthy patron, whom I have alfo the honour to call my friend, will do me the juftice to acknowledge that my Effay was in his hands, and that I had agreed with the bookfeller for the publishing of it before Mr. Crawford's, or even Dr. Leflie's performance appeared.

I AM weak enough to confefs that I fhould like to have had a fhare in the honour of this difcovery, and perhaps the candid in the learned world will not refufe me fome credit on that head, on a review of the evidence before them. Yet I fhould not have publifhed this Effay after having read Mr. Crawford's treatife, but that it was already in the prefs, and that there are a few points in which I had gone perhaps farther than that gentleman, or in which he does not feem to be fufficiently clear.

THE general caufe of combuftion *, for ex-

---

* It is worth while to obferve that extraneous fire may be faid to heat bodies *pofitively*; phlogifton, friction, &c. *negatively*.

ample,

ample, as far as relates to heat, is elegantly
affigned by the learned author ; and the different proportions of fire which he has fhewn to
be contained in fixed and atmofpherical air, are
far greater than I imagined *a priori* *. The
light, the different colours of the flames, and
fome other particulars which he has not attended to, may perhaps be underftood, in fome
meafure, from my Effay.

" IT is probable," fays the ingenious author,
" that the vapour of pure nitrous acid contains
as much abfolute heat as atmofpherical air ; for
the power of the former in maintaining flame
is nearly as great as that of the latter. In the
deflagration of nitre, the acid is converted into
vapour, which being the fame moment combined with the phlogifton of the coal, the fire is

---

* Had I meafured the degrees of heat before, and during
the combuftion, and noted the quantity of air confumed, this
difference might in fome meafure, have been difcovered ? but
through want of conveniencies, I was obliged to content myfelf with a folution in the *grofs*. The difference of elafticity,
&c. in fixed and atmofpherieal air, I confefs, does not feem
to anfwer to their quantities of fire, as difcovered by Mr. Crawford; perhaps fome of the fuggeftions at the end of the VIIth.
and IXth. feftions will better account for it.

inftantly

inflantly difengaged, an elaftic fluid is generated, and a loud explofion produced." That *air* was really contained in the nitrous acid previous to the combuftion *, is evident from the fixable air which is generated. For if the vapour had been merely of an aqueous nature, though it might have been expanded by the heat, it would have been condenfed into an aqueous liquid, and not fixable air, when cold †: and this being admitted, the Author's fuppofition, that " the vapour of pure nitrous acid contains as much abfolute heat as atmofpherical air," will appear to be true, and is a farther confirmation of his excellent theory.

Mr. CRAWFORD has proved that phlogifton and fire are different fluids, contrary to what has hitherto been imagined. I had attempted to fhew that *æther* was a third fluid, different from both thefe.

OUR Author imagines the attractions of bodies for phlogifton to be proportional to the degrees of heat neceffary to begin their com-

* Vide fection VI.   † Vide fection VII.

O 2   buftion,

buftion, and I once fell' into the fame error ; but that this is not always the cafe will, I think, appear by what is faid on that fubject in the VIIIth. fection of my Effay. May not the differences there obferved be partly owing to the different ftates of the double affinity ?

I had endeavoured to fhew that phlogifton diminifhes the attraction of bodies for fire, in proportion to the force of its combination ; and that this force of combination is capable of being intended or remitted even in the fame body. But of the latter of thefe propofitions, Mr. Crawford does not feem to have been apprized.

That Gentleman, in one part of his admired performance, has run a comparifon between fire and fixed air ; but fixed air is not regulated by an equilibrium, or common temperature, like fire *, neither does it appear that fire exifts in a loofe

---

* Fire does not feem to cohere with, and form an effential part of bodies like phlogifton, and fixed air. If the phlogifton of a metal, or the fixed air of marble be taken from them, the nature and conftitution of thofe bodies are quite altered, or they are decompofed ; but this does not happen with regard

to

a loofe or feparate ftate like that fluid, but
is retained by the particles of bodies according
to a certain law, if my reafoning be juft. The
affections of the fenfe, and of the thermometer
by heat, are not, I think, conceivable but by
admitting that law. (Vide cafe xi. fect. VIII.)
When phlogifton is added to air, the fire, according to my idea, is not *expelled* *, or does
not *fly off* in the manner of fixed air from
bodies; it would ftill be retained by the air,
though in a lefs forcible manner, if there were
no other bodies near, or bodies by which it
was not more powerfully attracted. If the bodies around were hotter, it would even attract
fire from them; but this is fpoken with fub-
miffion to better judges,

In regard to animal heat, I find by our excellent Author's experiments, that I had fallen

to their fire, at leaft there does not feem to be the fame kind
of analogy as between phlogifton and fixed air. There feems
to be a greater analogy between electricity and fire in this
refpect.

* I have ufed this word in many places, but in fuch a manner as to carry with it the above meaning.

into

into an error in imagining that the blood is partly heated in the lungs *. *That heat however is generated by the decompofition of air in that organ,* appears from hence, that the air which is expired is hotter than that which is infpired; and alfo by the following quotation from our Author's work. " By the heat of the furrounding medium, the evaporation from the lungs is increafed. Now it may be fhewn, that if the evaporation from the lungs be increafed to a certain degree, the whole heat which is feparated from the air will be abforbed by the aqueous vapour." And by the converfe, if the evaporation from the lungs be diminifhed, the

* Mr. Crawford in his firft propofition affirms, that air is fitter for refpiration in proportion to the abfolute heat which it contains. But it ought to be obferved that air is rendered unfit for refpiration by other means befides phlogiftication; as by particles floating in it which irritate the lungs, and the like: thus, in combuftion of fulphur, the acid is as prejudicial in this refpect as the phlogifton. This confideration does not feem to have been fufficiently attended to of late; but afthmatic people daily experience the juftnefs of the remark. Pure fixable air does not kill by irritating the lungs, but by not carrying off the phlogifton of the blood. The falubrity of air therefore cannot be determined by the eudiometer alone with fufficient accuracy.

blood

blood will be heated. But in general the evaporation from the lungs is fo proportioned, that
the heat of the blood is not increafed in paffing
through that organ : my error arofe from not
attending to this circumftance, which indeed
could not have been known but by experiment.

THE only inftance, of moment, of our admirable philofopher giving into hypothefis is
with refpect to the origin of the phlogifton im-
bibed by the blood; he fuppofes that it is taken
" from the putrefcent parts of the fyftem."
But I fee no reafon why the phlogifton from
thefe parts may not be difcharged either
wholly, or chiefly; by perfpiration, and by
urine ; and that it is fo, feems apparent by
the very great quantities which thefe excre-
ments, efpecially the former, contain. Neither
does the neceffity of the elaborate, and (to all
appearance) very important procefs * which
he had been defcribing†, on this fuppofition, ap_

* Refpiration, &c.
† Hence Mr. Crawford is at a lofs when he comes to apply his principles to cold animals.

O 4　　　　　　pear.

pear. I may have been prepoſſeſſed in favour of my own hypotheſis, and therefore may not be a proper judge of this matter : as a proof of it, I ſhall not ſcruple to confeſs that I felt myſelf pleaſed on finding that Mr. Crawford, in this part of his work, deviated from my track ; I ſhall therefore leave the merits of the two hypotheſes to be determined by the more impartial Reader, or by Mr. Crawford himſelf; for I have conceived ſo good an opinion both of his judgment and candour, from his admirable performance, that I would cheerfully acquieſce in his determination.

Having taken the freedom to point out another's errors, I ſhould next proceed to enumerate my own. The taſk, however, would now not only be laborious, but uſeleſs, as my theory was only given by way of conjecture; though, for the contrary reaſon, it was proper to notice any error of Mr. Crawford. Some of my miſtakes I have already mentioned ; the following, though not ſhewn to be ſuch by Mr. Crawford's experiments, were yet diſcovered by a more ſtrict review of the foregoing Eſſay in
conſequence

confequence of that Gentleman's very ingenious publication.

In the VIIth. fection I adopted Dr. Black's fuppofition, that *phlogifton has a centrifugal tendency*, and thought I had accounted for it by imagining that its particles attracted æther. But though this fhould be allowed, yet unlefs the globe of the earth did alfo, there would be no repulfion between them. Particles of earth gravitate becaufe æther is mutually repelled by them, and by the terreftrial globe ; and, there‑ fore, if phlogifton attracts æther, there can, at moft, only an indifference be produced in it with refpect to gravity or levity, the globe of the earth, and the particles of phlogifton mu‑ tually deftroying each others effects. The rays of light, fetting afide their *inflection*, which may be otherwife accounted for, do not feem to have either centripetal or centrifugal tendency ; or if they have either it does not appear to be in any confiderable degree: the like may be obferved of electricity, which I take to be phlo‑ gifton in the next degree of purity to light. If fire attracts æther, fire alfo muft be alike indif‑

ferent

ferent with regard to gravity or levity, and this feems, by experience, to be the cafe. There are however, methods of conceiving how phlogifton may diminifh the gravity of bodies; or do metals, &c. attract air, when calcined, in lieu of their phlogifton, and thereby have their weight increafed? for an effervefcence attends the reduction of thofe calxes *, as I find by the Chymical Dictionary: fhould this latter be the cafe, it would appear (and it would be a little extraordinary) that *error* had led to me *truth.*

THE above, as it has accidentally been the fource of right, fo it has alfo of wrong reafoning, as may be feen in the courfe of the VIIth. fection. I do not however find reafon to reject the general fyftem fuggefted in that fection; and am ftill inclined to think that *Æther, Fire, Phlogifton,* and *Earth* are the four principles of

---

* Dr. Prieftley fhews that dephlogifticated air may be expelled from the calx of lead. The dephlogifticated air may have been formed of fixed air which the calx had attracted, and afterwards decompofed. (Vide § VI.)

which

which the world is compofed, (taking alfo into
confideration what was faid of their properties
in page 127): but the following feems to be a
more proper arrangement of them, and of their
ufes. Æther, and earth, are mutually repulfive;
hence the gravitation of the particles of the
latter: fire and phlogifton are principles intermediate to thofe; the latter feems to be
the principle of cohefion among the particles
of earth; the former of their feparation.
Phlogifton is alfo the caufe of light, fire of
heat, and on the various compofitions or affo-
ciations of the above principles, the fenfible
qualities of bodies, and the phænomona of nature in general, feem to depend.

As I formally renounce the falfe reafonings
which may be met with in the VIIth. fection,
and fome other parts of this Effay, candour, I
prefume, will prevent their being brought in
judgment againft me.

IF any confiderable part of the foregoing
work fhould have the good fortune to be approved, the errors which I have difcovered, and
any

any others which may in the mean time appear, would be omitted in a future edition. They had certainly been fo in this, as well on my own account as the Reader's, had I feen Mr. Crawford's performance in time. The greateft philofophers that ever lived have fallen into errors *, efpecially where experiments where wanting to afcertain the truth; it is no difgrace to err in fuch good company, and as I make the critic my *prieft* by confeffing to him my *faults*, I have a firm *faith* in his fupporting the chriftianity of the character by granting me *abfolution*.

* The recent inftance of Dr. Leflie on the fame fubject, a gentleman who had every advantage over me in point of information, might be urged; and alfo the opinion of the gentlemen concerned in the Reviews on his performance.

P. S. It

P. S. Does not Dr. Prieftley's difcovery of
" light decompofing fixed air in water." depend
on the principle laid down in the XVth. cafe,
and applied to friction, percuffion, &c.* at the
end of the IXth. fection? are there not in the
water particles, either of the water itfelf, or
of more fixed fubftances, which have originally ftronger attractions for phlogifton, than
the particles of air? do not the rays of light,
by their action on thofe particles, increafe
their attractions for phlogifton, and thereby
enable them to take it from the particles of air
contiguous or perhaps in combination with
them, agreeable to the principles above alluded
to? hence heat has not this effect. It feems
therefore to be analogous, in principle, to the
decompofition of fixed air by agitation with
water, a former difcovery of that excellent
philofopher.

I HAVE fuppofed in page 195, that the phlogifton is derived to the nerves, not from the

* That is, to the putting the particles of bodies into vibrations; though the heat is not the immediate effect of thefe
vibrations, as has been imagined, but as explained in fection IX.

brain,

brain, but from the chyle in the various parts
of the body ; as the chyle is agitated in and
againſt the veſſels, may not this be brought
about on the ſame principle? Dr. Leſlie advanced a propoſition, that " by the action of the
veſſels, the phlogiſton of the chyle is gradually
evolved throughout the body." There ſeems to
be ſome truth in the opinion, though that
Gentleman, imagining phlogiſton and fire to be
the ſame, erred in the conſequences which he
drew from it. If the above be the true ſtate of
the caſe, and if what is advanced in the three
laſt ſections of my Eſſay be juſt, the phlogiſton
is transferred from the chyle to the nerves, from
the nerves to the fibres, from the fibres to the
blood, and from the blood to the air. Some
other operations of nature may probably depend
on the ſame principle.

T H E E N D.

·